ILLUSTRATING FOR SCIENCE

$$\frac{38}{\sin\theta_W}\ \text{GeV}$$

WATSON-GUPTILL PUBLICATIONS
NEW YORK

Editor: Paul Lukas

Senior Editor: Marian Appellof

Designer: Jay Anning

Graphic Production: Ellen Greene

First published in 1992 by Watson-Guptill Publications,

a division of BPI Communications, Inc.,

1515 Broadway, New York, NY 10036

Library of Congress Cataloging-in-Publication Data

Kelvin, George V.
     Illustrating for science : A problem-solving approach to rendering subjects
   in biology, chemistry, physics, astronomy, space technology, medicine,
   geology and architecture / George V. Kelvin
       p.   cm.
   Includes index.
   ISBN 0-8230-2540-3
   1. Scientific illustration.   I. Title.
Q222.K45 1992
502.2`2--dc20   91-34136
                                                      CIP

Manufactured in Singapore

First printing, 1992

1 2 3 4 5 6 7 8 9 / 98 97 96 95 94 93 92

For Deanna, with love

And for Roy, Ward, and Rochelle, for the happiness they add to my life

And in memory of my mother, who encouraged and supported me
through my early art training and bought me my first airbrush

ACKNOWLEDGMENTS

I would like to express my appreciation to my friend and editor Mary Suffudy, who knew of my work as a scientific illustrator, urged me to write a book on the subject, and guided me through its development. I should also like to thank Marian Appellof and Jay Anning, for their excellent work in designing and laying out the book, and Paul Lukas, whose suggestions and deft editorial hand are reflected throughout these pages.

I am indebted to the following colleagues and friends for supplying invaluable advice, research, and technical assistance:

- *Scientific American:* Dennis Flanagan, Jonathan Piel, Sam Howard, Joan Starwood, and Edward Bell

- W. H. Freeman Co.: Linda Chaput and William Page

- Saunders College Publishing: John Von Delling, Christine Schueler, and Carol Bleistine

- *Discover:* Paul Hoffman

- Time Inc.: Edward Hamilton and Arnold Holeywell

- W. W. Norton Co.: Joe Wisnovsky

- Capital Engraving: Bob De Huete

Thanks are also due to James D. Watson, Ph.D.; Robert Nenner, M.D.; Richard Winston, D.D.S.; Bernard Ambinder, D.D.S.; Arnold King, M.D.; Stanley Strick, M.D.; Alan Mazel, D.D.S.; Lawrence Rose; Mandel Dalis; Leonard Weiss; Harvey Barer; Edward Haberman; and Ted Baer.

Finally, a note of thanks to my wife, Deanna, and to our wonderful children, Roy, Ward, and Rochelle, whose continuous help and encouragement made this book possible.

CONTENTS

CHAPTER ONE

INTRODUCTION TO SCIENTIFIC ILLUSTRATION

If you are an artist or are studying to be one, this book will introduce you to the challenges and rewards of scientific illustration. For the intellectually curious artist, this field offers excitement and stimulation on two levels: Not only will your artistic skills and imagination be utilized to their limits, but you will be exposed to information at the cutting edge of technology. The opportunities to learn are boundless.

This book is designed as a style guide, a reference source for tools, materials, and techniques, and a handbook of shortcuts, secrets, and tricks of the trade. In addition to the book's obvious utility to artists, authors and editors in the sciences will find it to be an invaluable reference guide for solving illustration problems and for assisting them in developing the technical proficiency essential to good author–artist communications.

We live In a technologically sophisticated society, surrounded by an ever-increasing number of mechanical and electronic devices in the work place and the home. The comforts of modern living and the advances made in science, medicine, and technology in the last half-century have become as integral a part of our lives as the food we eat and the water we drink.

Consider for the moment the extraordinary variety of products available in today's marketplace—each of these had to be conceived, then designed and crafted, and finally manufactured. At every step, some type of illustration was required. Illustration communicates the visions of the inventor, the designer, and the researcher, while providing necessary guidance to the engineer, the machinist, the investor, and the financial backer. With the prospect of an expanding marketplace clearly in our future, the need for scientific and technical illustration can be expected to increase in direct proportion to the advances being made in science and industry.

A faint glow surrounds the Challenger space shuttle as it speeds through an atmosphere of atomic oxygen and molecular nitrogen 250 kilometers above the Earth's surface.

BACKGROUND AND TRAINING

I grew up in Brooklyn, New York, in a neighborhood with endless opportunities to learn from experience. Toys were expensive, but there was a wealth of raw material around to satisfy my curiosity. A visit to the grocery store yielded smooth wooden cheese boxes with finger-notched joints, the kind they don't make anymore. And the tailor down the street could be counted upon to save me a few hangers and wooden spools from thread.

A pocket knife was all I needed to convert a cheese box into a housing, which I fitted with pulleys, axles, a wire crank, and a rubber band. With various modifications, this crude machine was a windmill one week and a rubber band–powered boat the next.

Later on, hobby magazines, with their "build it yourself" projects, became a marvelous source of entertainment, giving me my first opportunity to work from technical instructions, which I often supplemented with my own sketches. In high school mechanical drafting class, I moved on to blueprints and learned to apply basic geometric principles to complex forms and structures, which would prove invaluable on various professional projects years later.

As my education progressed, I developed an interest in the natural sciences and particularly enjoyed art-related projects. I was struck by the similarities between the organic and mechanical worlds—water moving up a plant's stem to the leaves, for example, is very much like the transfer of energy from a clock's spring mechanism to the gears and hands.

My artistic skills actually began to surface while I was in the Army, where I was in charge of posters and field-training graphics for my base. Following my discharge, I enrolled in art school at Pratt Institute, where classes were six hours a day, five days a week. Then, as now, the foundation year provided the most important training, with daily classes in figure drawing, anatomy, color, two- and three-dimensional design, and perspective; subsequent semesters included advanced courses in design, painting, and illustration.

My purpose in recalling this early training is to stress the need for a strong foundation in the formative years. Career goals should not be the overriding consideration in structuring a course of study; building a broad-based range of skills is far more important. In addition to artistic ability, mechanical or technical interests are desirable if you are thinking about illustrating for the sciences.

Some artists working in the field manage to get by with only minimal technical or scientific training. Their illustrations reflect this, often amounting to little more than elaborate renderings of reference material. Moreover, their inability to contribute to or even assess the technical accuracy of their own work tends to relegate them to the lowest rung of the professional ladder—the most prestigious assignments are offered only to the artists who demonstrate a sufficient grasp of the material.

Abstract interpretation of molecular structures.

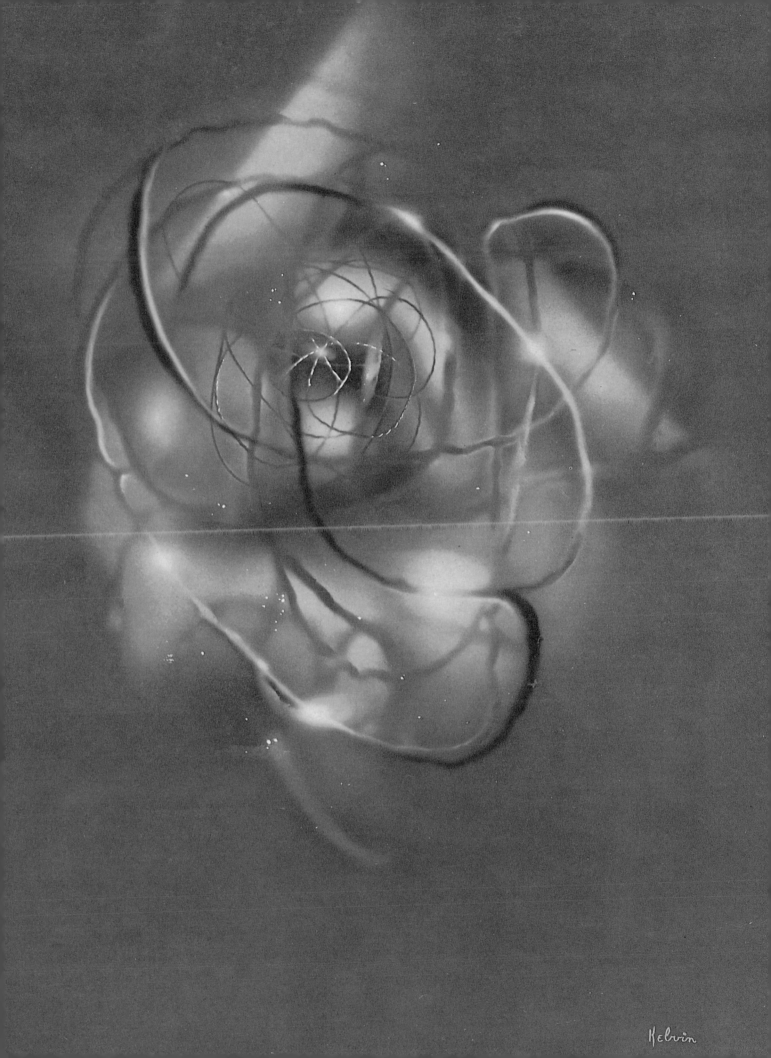

THE ROLE OF THE SCIENTIFIC ILLUSTRATOR

In today's complicated world, difficult concepts have to be simplified and explained by means of illustrations. These renderings are fundamental to the accurate dissemination of complex information in every area of science and technology. Viewed in this context, the scientific illustrator plays a vital role in modern society, communicating the increasingly complicated underpinnings of everyday life in comprehensible terms. This role requires a lifetime commitment to learning, a dedication to accuracy, and a focused understanding of geometry and spatial relationships.

With most shapes in nature based on geometric forms, the knowledge acquired in one area is generally applicable to others. The same geometric principles needed to represent structures in molecular biology, for example, are also essential to drawing a crystal structure, a virus, a galaxy, or a subatomic particle. However, while the principles may remain the same, the geometry may be many times more complex.

The scientific illustrator, therefore, must be a good researcher. Careful observation and accurate measurements are the basis of good research—just as a trained detective gathers clues, a skilled illustrator learns to extract information from every available reference source. Not until all the data has been assembled can the artist determine which direction to take.

Using this approach, design problems should always be worked out in a preliminary sketch. If the initial pencil rendering looks like a convoluted maze, the final art will not be any better.

Thirty-six electron beams concentrate upon a deuterium-tritium fuel pellet, imploding it, in an experimental electron-beam fusion accelerator.

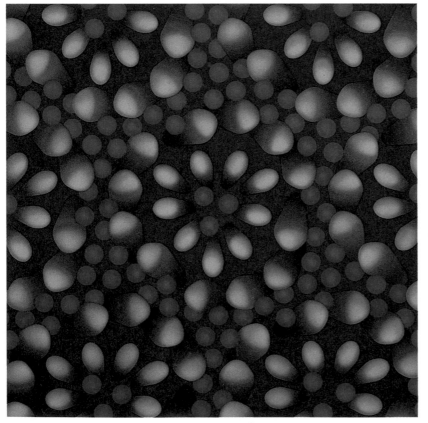

A three-dimensional top view of bateriorhodopsin and lipid molecules of a cell membrane. The repeating units form a geometric crystalline array.

Molecular lattice structure of a proposed microchip. The green pathways represent the current flowing through the open switches, which are indicated by white dots.

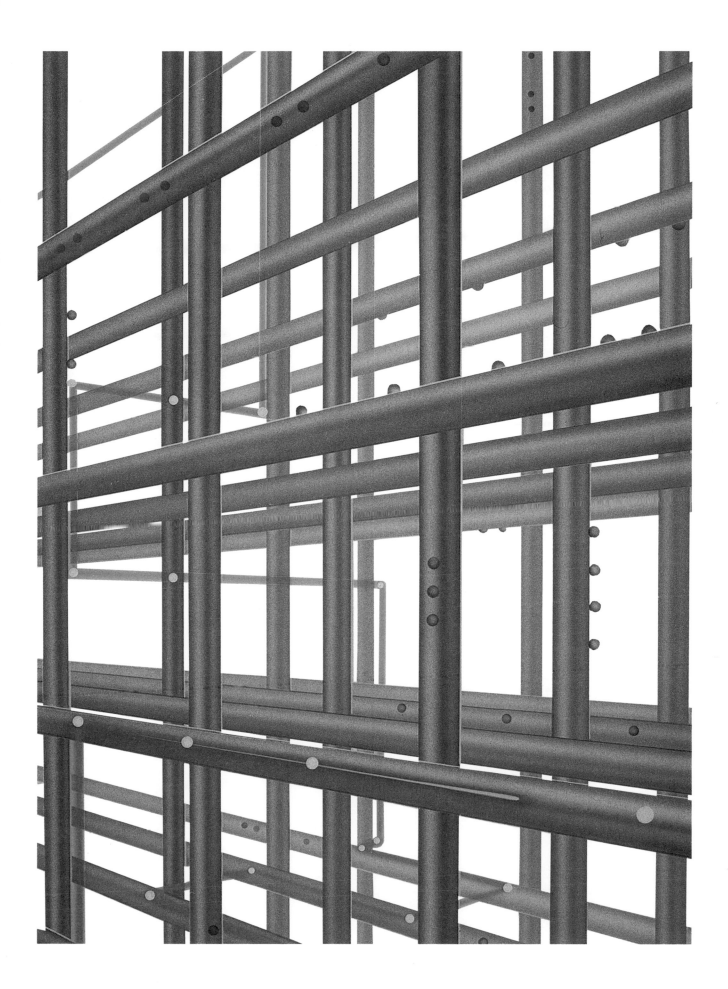

CLARIFYING INFORMATION: SIMPLIFY, SIMPLIFY, SIMPLIFY

The science illustrator must always keep in mind the objective of a given assignment. For example, a drawing meant to show an object's structure and design is handled quite differently than one meant to show a basic principle of operation. These two goals often overlap, requiring art that is simultaneously representational and descriptive. How much weight to give each element is, of course, a professional judgment call that must be based on many factors, including intended function, accompanying text, and target audience.

One of the hardest decisions is choosing between presenting a complete set of information or simplifying the art for the sake of clarity. An illustration designed for a professional audience, for example, may involve a variety of very complex information. If the components are so complicated that they interfere with the illustration's communicative function, the artist must simplify the concepts.

One method of dealing with this problem is by breaking down the illustration into two or more steps and modifying the different concepts wherever possible. Color coding is a useful adjunct when portions of the illustration are being referred to in the text. Adjustments in scale, noted for the audience in a legend, can be imposed when size differentials are so great that drawing one part exactly to scale would cause another to be larger than the page. Adjustments may also be called for if one object is eclipsed by another, thereby preventing it from being seen or identified.

An example of this type of artistic license can be found on any topographical map featuring mountain ranges. Typically, the horizontal scale is the true scale, while the vertical scale might be shown at up to ten times its actual height. While this is a distortion of fact, it makes it possible to produce a graphic representation of features and configurations that otherwise would not be possible.

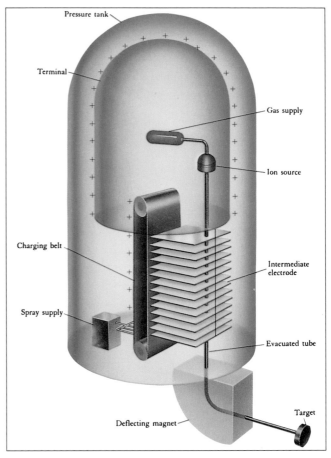

Simplified depiction of a Van de Graaf generator, intended for readers of a general science book. Most of the support structures were eliminated to make the operational procedure easier to follow.

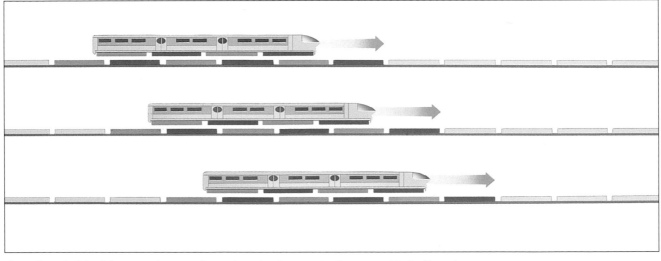

Operating principle of the magnetic suspension and acceleration aspects of a proposed train. The red rectangles in the track represent superconducting magnets in the "on" position, and the three steps of the illustration show the progression of the magnetic pulse and its propulsive effect on the train.

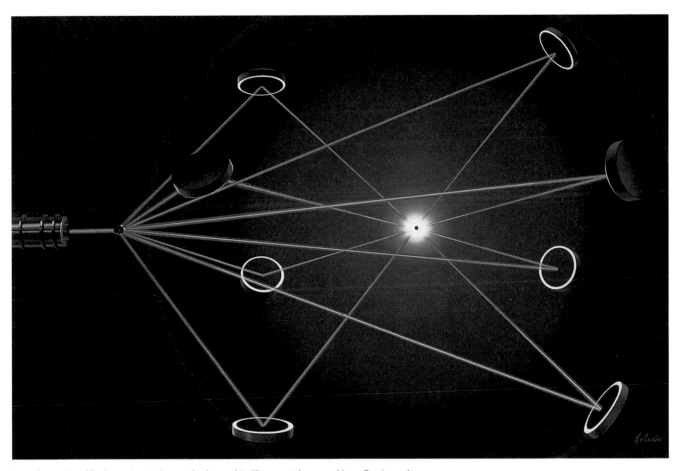

In order to simplify the presentation, only the multitrillion-watt laser and its reflecting mirrors are shown in this nuclear fusion illustration. All extraneous structures have been eliminated.

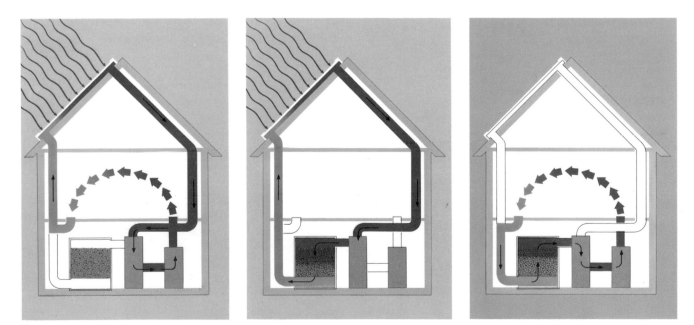

A solar heating system, shown in three modes of operation. The simple, uncluttered shapes make the concept easy to understand.

Left: Radiant solar energy (represented by red wavy lines) warms the air in the rooftop energy collector (red). From there it is drawn down into the basement, where an air controller (tan) circulates it into the house. As the air gives up its heat, it is recycled to the energy collector on the roof (blue). A conventional auxiliary heater (tan) helps to boost the heat on cloudy days.

Center: Hot air is diverted to a basement bin of crushed rocks, where heat is stored until needed.

Right: At night (indicated by a blue background), the heat stored in the rock bin is circulated through the house.

SIMPLIFYING A COMPLEX SYSTEM

Particularly complex concepts and structures can sometimes seem impossible to simplify. What can be altered or modified? What can be shown at an altered scale? What should be stressed or emphasized, and at the expense of which other elements? Working from a model can lead to a solution.

Shown here is an extremely modified rendering of a laser, designed to give readers of a general-interest magazine an approximate idea of its operation. A real laser has glass tubing wound in a tight spiral around a glass sleeve; inside the sleeve is a ruby core. When the laser beam is activated, gas in the tubing generates a brilliant white glow. In order for the reader to follow the laser's operation as described in the text, I had to provide a clear view of the ruby core, which I accomplished by reducing the number of tubing coils and eliminating the white glow of the gas discharge.

Before starting the illustration, I constructed a model of the laser as a drawing guide. I formed the cylindrical housing by rolling a sheet of clear acetate to a diameter of 1.5 inches (3.8 centimeters), cut to a 16-inch (40.6-centimeter) length, and then carefully wound a rubber tube around it, creating a loose spiral—a very simple but effective procedure. Using a Polaroid camera, I took several shots, changing the viewing angle and spacing of the spiral wrap with each shot. With the pictures in hand as reference, I decided on the view that most clearly showed the operation and the structure.

The time involved in constructing the model, taking the pictures, and selecting the final photograph was about half an hour. After this, a working-size enlargement of the photo became the foundation for a preliminary sketch. Correcting for minor flaws as I went, I made pencil tracing of the spiraled tube, adding the core and beam to the drawing. In order to observe the subtleties of refracted light and color playing on the glass surfaces, I placed a cylinder of red paper behind a glass tube and transferred what I saw to the final art.

Too many coils—the core is hidden.

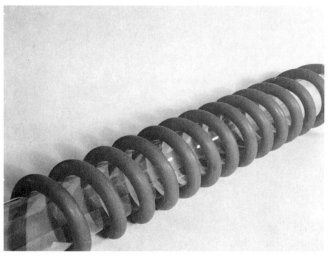

Changing the viewing angle did not alleviate the problem.

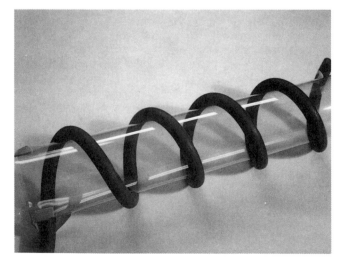

Loose coiling is an improvement, but the spiral is hard to follow.

This photo, taken from a sharper angle and ultimately used for the pencil layout, shows the continuous spiral I was after.

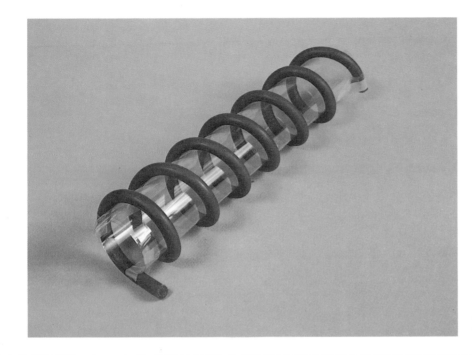

Left: Photograph of completed laser model.

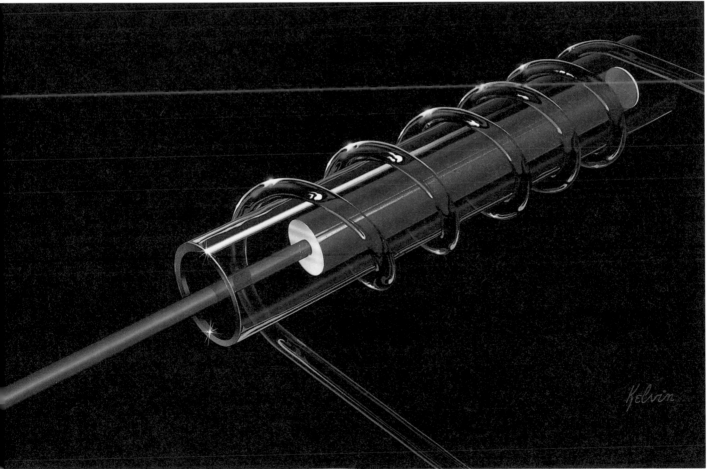

Final illustration.

VISUALIZING ABSTRACT CONCEPTS

Half the work that comes across my desk falls into the category of conceptual art. The rest comes with the reference material and documents needed to do the job, obviating any need for me to do the research myself. What still remains, of course, is making sure that the reference material checks out and is accurate and current.

Illustrations that are purely conceptual in nature, often dealing with forms and objects that have never actually been seen, are more demanding. Since solid reference data is not generally available for such projects, the art director, author, and artist must work together to establish reference guidelines for the illustration. This is done by establishing what is known, what is probable, what is unknown, and what can be assumed.

Gathering this information may temporarily require the illustrator to become a researcher. The best way to go about this is to consult an acknowledged expert on the subject, who will often turn out to be the author. Once the expert is located, the questions that should be asked are as follows:

1. How many component parts must be included?

2. How should these parts physically relate to each other and to the whole?

3. What sizes and configuration should the parts have?

4. Are there any specific color requirements?

5. Is there a scale problem that might prevent all the components from being shown in the same drawing?

6. How do the components interact with each other (mechanically, chemically, biologically, sequentially, mathematically)?

In addition, each discipline has its own set of questions to address before a conceptual drawing can be developed. The thorough artist will not begin until all such queries have been answered. For instance, if I were asked to produce an illustration of the Cadillac of the future—the year 2010, say—the experts at GM would have to tell me if the car would still have four wheels, whether it would be able to fly, and so forth.

When astronomers observed a star moving in an unusual manner, its back-and-forth pattern led them to hypothesize that there must be a companion object close by, with the two bodies exerting a gravitational pull on each other as they rotated about a common axis. The second object was thought to be a collapsed star, which would explain why it could not be seen, since collapsed stars do not emit light. This series of sketches was based on the available information.

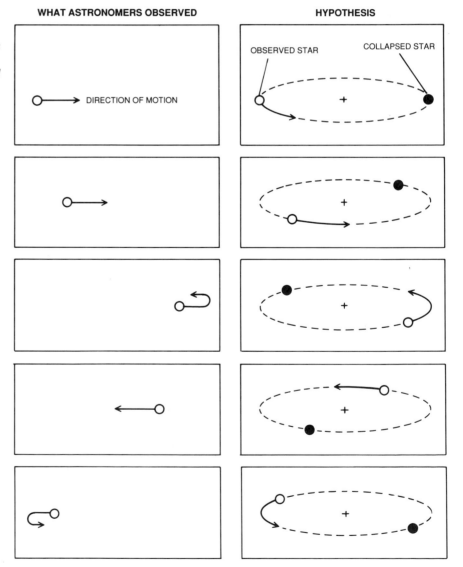

WHAT ASTRONOMERS OBSERVED

HYPOTHESIS

OBSERVED STAR   COLLAPSED STAR

DIRECTION OF MOTION

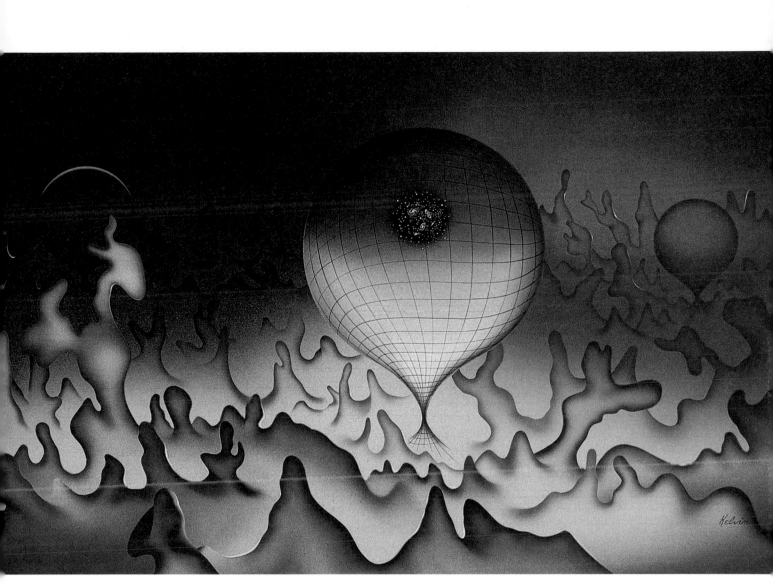

A much more sophisticated example is represented in this astronomical painting of a theoretical process of how galaxies form in an expanding universe. Some of the structures are based on observed data; others are purely hypothetical. Color was used to aid in identifying areas discussed in the text.

KEEPING YOUR AUDIENCE IN MIND

An illustration should be planned with its eventual audience in mind. Science textbooks for elementary school students, for example, address fundamental concepts and should feature easily followed illustrations, some of which might be designed to read like a story, with a beginning, middle, and end. Such stories should contain well-defined main characters and a less intrusive supporting cast, with particular attention given to sizing and positioning of the various elements, along with careful color selection.

At the high school level, diagrams and illustrations should show a more complete representation of information, with labels and legends employed for the more complex descriptions. At the college level, illustrations and text become quite sophisticated and demand detailed drawings encompassing an extensive range of information.

The continued emergence of new outlets for illustrative work has changed the role of the illustrator, who now needs to know the basics of many disciplines in order to keep pace with the ever-increasing range of opportunities in media, science, and industry. The primary parameters of the field, however, remain: On one hand, scientific illustration depends on traditional art skills and techniques; on the other, it is important to think ahead and work to expand every aspect of one's knowledge and training.

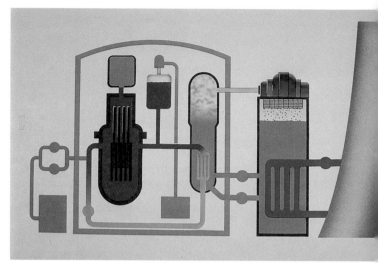

For an elementary school representation of an atomic power plant, the concepts are kept simple and diagrammatic. A simplified silhouette is employed; scales have been altered for clarity.

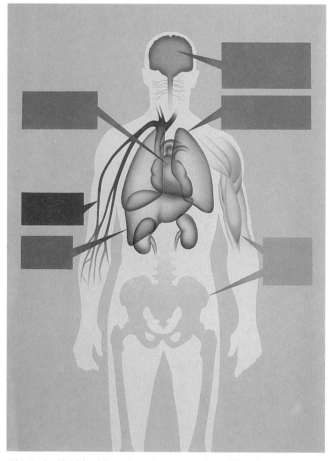

This is the "laid-back" version of a more realistic "blood-and-guts" depiction that would normally appear in a university- or professional-level book. The editors stressed the importance of presenting pleasant, nonthreatening illustrations for young students just being introduced to the sciences. The use of soft, muted colors and a neutral background met this goal. Labels identifying organs, muscles, and bones were added to the color-coded boxes.

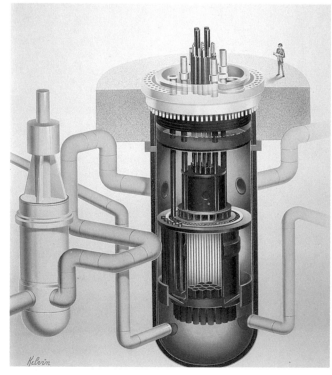

For high school students and readers of general-interest magazines, the plant is shown in greater detail. The structures are realistically drawn and the scale is correct.

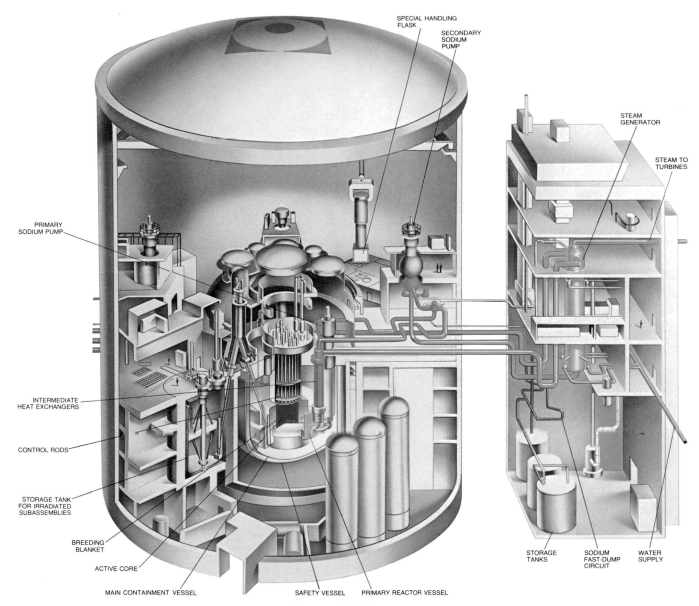

PRIMARY SODIUM PUMP

SPECIAL HANDLING FLASK

SECONDARY SODIUM PUMP

STEAM GENERATOR

STEAM TO TURBINES

INTERMEDIATE HEAT EXCHANGERS

CONTROL RODS

STORAGE TANK FOR IRRADIATED SUBASSEMBLIES

BREEDING BLANKET

ACTIVE CORE

MAIN CONTAINMENT VESSEL

SAFETY VESSEL

PRIMARY REACTOR VESSEL

STORAGE TANKS

SODIUM FAST-DUMP CIRCUIT

WATER SUPPLY

At the college and professional levels, the operation's complexity is clearly shown. Careful attention is given to scale, components are extensively labeled, and color coding identifies the various systems.

CHAPTER TWO

PREPARING ILLUSTRATION FOR PRINT REPRODUCTION

There is a wide range of tools and equipment from which the scientific illustrator can choose. Some of the brushes and paints are so basic that they date back to the earliest years of the graphic arts. Others, like the computer, are so revolutionary that we are just beginning to explore their possibilities.

Every illustrator's approach to solving problems is dictated to some extent by his or her choice of tools and equipment. These choices are very personal decisions—no two artists work quite the same way. You may want to experiment with some of the high-tech equipment and precision tools that have become available, but only invest in those of the highest quality. As for consumable art supplies, I suggest limiting the selection to those that have proven to be the most versatile and reliable for reproduction.

This point regarding reproduction is worth emphasizing. Unlike the fine artist, whose work may hang in galleries or on living room walls, the scientific illustrator must keep in mind the ultimate goal of *print reproduction*. Everything discussed in this chapter is directed toward achieving optimum results in that area.

Two months' worth of acetate masks from completed jobs.

STUDIO SETUP

A professionally equipped and sensibly organized studio is essential for any working illustrator—without a secure, well-stocked place to work, nobody can expect to turn out consistently high-quality work. An examination of my own working environment may give others a clearer understanding of their own needs in this area.

My studio occupies the third floor of my house. There are several distinct advantages to working at home, including a short and convenient walk to the office and considerable flexibility in arranging my working day. The studio space is approximately 16 by 26 feet (5 by 8 meters), loosely divided into specialized areas designed for maximum efficiency. Three cool-white fluorescent ceiling fixtures, each 8 feet (2.4 meters) long, provide bright, evenly distributed light throughout the room, creating a comfortable working environment and consistent illumination that is unaffected by the day-to-day variations of natural light.

In the center of the room is the primary work space: My drawing table, measuring a generous 4 by 6 feet (1.2 by 1.8 meters), is equipped with a drafting machine and two additional fluorescent lamps, which throw a shadowless light over the entire table area. Three airbrushes hang from hooks along the table's edge. The chair is ergonomically designed and fully adjustable, an expensive but necessary item for the illustrator who wishes to avoid back problems.

I am ambidextrous, drawing and painting with my left hand but taking notes with my right, and my studio is set up to accommodate this. As I sit at the drawing table, there is a large formica shelf to my left, where my paints, adhesives, and assorted supplies are laid out. The cabinet below holds tracing paper, acetate, and illustration and mounting boards.

Notes and reference material pertaining to work in progress hang above the cabinet. To the right is a 12-drawer taboret, giving me easy access to an array of pens, brushes, triangles, curves, and other drawing instruments.

Behind my desk is a large, all-purpose table, with a paper cutter at one end and a copy machine at the other. On the other side of the studio, a 35mm camera is set up for taking color slides of final art. When this camera is in use, the overhead fluorescents are turned off, so as not to affect the sensitive color balance of the transparencies, and four 500-watt, 3,200K bulbs in reflectors mounted on a ceiling-hung light bar are used instead. When I am photographing reference models in black and white, it is convenient to set up right on my drawing table and use a Polaroid camera and the table-mounted fluorescent lamps.

My fax machine has turned out to be one of the more time-saving devices in my studio—it puts me in any office, anywhere, instantly. I use it primarily to send sketches back and forth for on-the-spot corrections.

At the far end of the room is the business office, a self-contained area with its own desk, chair, and telephone. Along the back wall is my reference library, which holds a very select collection of technical and scientific material and gives me access to a valuable source of ready information and reference art.

A large metal cabinet with five full-extension compartmentalized drawers holds my collection of assorted mechanical parts, pieces, and odds and ends—everyone needs one of these. Other cabinets, vertical files, storage space, and a huge philodendron fill the remaining nooks and crannies around the room.

Art table setup.

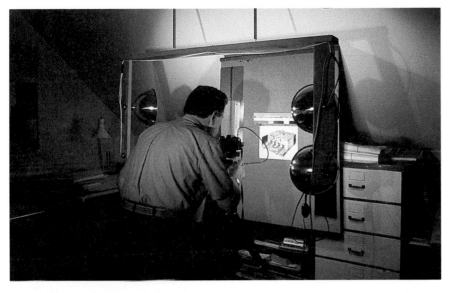

Setup for photographing final art.

Setup for close-up detail photography.

Copying machine.

Maintaining a business office within the studio provides the best approach to monitoring supplies, bills, expenses, and so on.

A basic reference library is essential for any working scientific illustrator.

Storage files and auxiliary layout/model desk.

Vertical storage file.

ESSENTIAL MATERIALS

ART BOARD

Because all of my work is designed for reproduction in a print medium, it is essential that my painting surface be ideally suited for my airbrush technique and subsequent photo-engraving. With most color separation work now being done by laser scanners, art boards must be pliable enough to be wrapped around the scanner's rotary drum yet stiff enough not to buckle or crease. Professional-quality Strathmore white four-ply bristol board meets all of the necessary requirements. The board has a medium surface with a slight tooth, and is excellent for a wide variety of techniques in addition to airbrush application.

PAINTS FOR REPRODUCTION

Winsor & Newton Designer's Gouache paints are excellent for airbrush work and reproduce faithfully. While watercolors and dyes will produce original art with a beautiful luminous quality, laser scanners frequently overreact to the white of the paper that shows through these paints. The typical result is a partial burnout of color, causing the printed illustration to look flat, dull, and washed out. The same holds true for very thin washes of acrylic paint.

Keep in mind that the wavelengths of fluorescent colors cannot be detected by a laser scanner; such paints, therefore, should never be used on art intended for reproduction.

BLACK INKS

There is a wide variety of black inks from which to choose, some formulated for use on paper, some for mylar or acetate. Some are permanent, others water-soluble or erasable. Use the ink that is most appropriate for your surface.

Most inks start to thicken after a year or so, and in some cases the pigment separates and sinks to the bottom of the bottle, leaving just a thin, watery solution above. With this in mind, I throw out all my ink and buy a new supply at the beginning of each year.

Waterproof inks have a greater density than their nonwaterproof counterparts, making the former a better choice for reproduction purposes. In my studio, pen and ink work is always done with a Rapidograph pen and Koh-i-noor 3080-F waterproof ink on mylar overlays. If art is needed for color transparencies, line work is done on prepared clear acetate, so that the base art and the overlay can be photographed simultaneously.

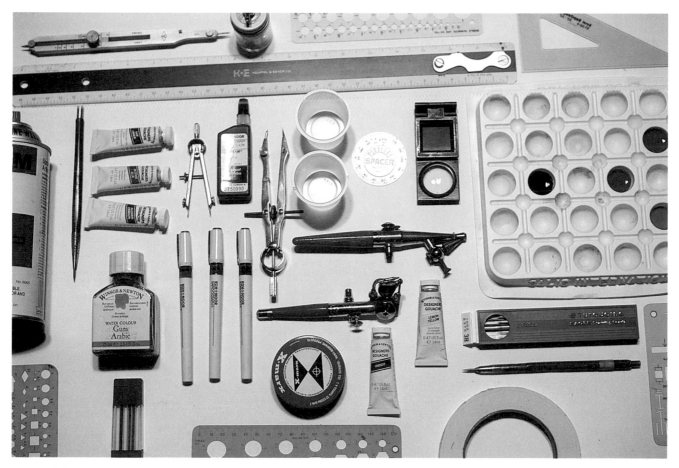

Successful illustrations begin with high-quality tools and materials.

THE AIRBRUSH

Unlike classical illustration, in which form and shape can be loosely suggested by a dab of paint, the precise nature of scientific illustration demands careful execution and delineation. I find the airbrush ideally suited to this task, producing an extraordinary level of photographic realism. However, the airbrush has more limited uses when illustrating for the natural sciences, such as biology, botany, and zoology. Birds, insects, animals, and other living subjects often require meticulous renderings by artists dedicated exclusively to these specialties.

Deciding whether a subject should be rendered with airbrush or paint and brush depends on the desired effect—stylized, photographic, realistic, and so on. In any event, the airbrush is the most versatile tool available and is unsurpassed as a production tool—in a matter of minutes, a skilled airbrush artist can produce flawless tonal gradations that would require hours of tedious blending with traditional brush technique. It also proves especially valuable when rendering repetitive forms and shapes and for achieving controlled stipple and blending not possible via any other method.

Skilled pencil rendering technique is an absolute prerequisite for proper airbrush execution—the entire technique depends on drawing ability. The tool alone does not make the artist, and anyone who cannot produce faithful pencil renderings of objects like a shoe, a wine glass, or a wrinkled handkerchief will fare no better with an airbrush. For the accomplished illustrator, however, the airbrush becomes another instrument of expression, taking its place alongside the pencil, pen, and paintbrush.

Airbrush art had its heyday during the 1940s and 1950s, when ad agencies relied on airbrush renderings to make the latest cars rolling out of Detroit look long, sleek, and shiny. My own introduction to the airbrush came during this same period, while working for a firm specializing in Army and Navy training manuals. In the 1960s, however, as high-speed, fine-grain color film became readily available and relatively inexpensive, photography began to replace much of the more costly airbrush work. Not until the mid-1970s was the airbrush again widely recognized as the valuable and versatile instrument it is.

Today's renewed interest in airbrush art is evidenced by the increasing number of pages art-supply catalogs now devote to airbrushes, compressors, specially formulated paints, and related equipment. Thanks to this resurgence, there is a wide selection of airbrushes on the market. I favor the Paasche AB for work requiring fine to moderate detail. It is designed with a reciprocating needle, driven by an air turbine that rotates at 20,000 revolutions per minute. Each time the needle travels back and forth, a tiny drop of paint is carried from the reservoir to the needle's tip, where it is blown off by a jet of air, creating a fine, continuous tone of color. Once the instrument's multitude of adjustments have been mastered, a broad range of subtle effects become possible.

The Paasche H is essentially a spray gun for painting backgrounds. The paint is held in a cup and drawn through a tube into a nozzle, where it mixes with compressed air and is blown onto the art surface. This tool has only two adjustments—one regulating the nozzle spray from narrow to wide, the other offering an air-flow range from fine to coarse to spatter. Three nozzle tips are available: Nos. 1 (fine), 3 (medium), and 5 (broad coverage).

Air is supplied to the airbrush through an electric compressor. If possible, choose one with a tank reservoir, which will keep the air pressure from varying. Compressed air is also available in aerosol cans, which effectively makes your airbrush portable.

Trouble-free operation and performance of any airbrush depends on conscientious maintenance. All parts have to be kept scrupulously clean and free of dried paint. After each color change, I run a few drops of clean water through the brush until the water runs clear. The various airbrush manufacturers give their own instructions for more specialized maintenance of their respective products.

PAINT PALETTE

An ideal companion to an airbrush is a 30-well plastic paint palette. These are generally sturdy enough to be washed out and reused many times in the course of a job, and also inexpensive enough to allow a new one to be used for each job.

My practice is to write the client's name on the side of the palette, along with the illustration reference number and any special instructions regarding paint colors. The palette is filed away with the project's job folder until the illustration is printed, eliminating the need to remix and match colors should the illustration require revision in the future. If last-minute changes are needed, the colors can easily be reconstituted with a little water.

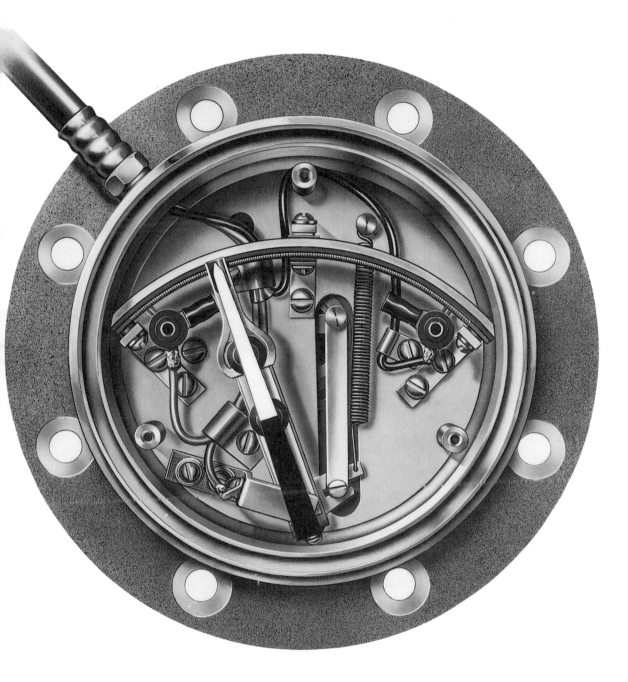

These airbrush renderings of a remote-reading fuel gauge (above) and the color spectrum (right) demonstrate the wide range of effects offered by this versatile tool.

FORMS OF ILLUSTRATION— LINE, TONE, AND COLOR

Illustration for reproduction can take many forms, and there are no hard and fast rules regarding which illustration forms are preferred for any particular subject matter. Indeed, the choice of art form is often determined by budget, time, and the artist's imagination. In all cases, however, the overriding consideration should be how well the art will reproduce in print. So the more the artist knows about print processes and preparing art for reproduction, the better equipped he or she will be to solve illustration problems in all the various art forms.

All original art falls into one of two categories: line and tone. Reproducing an original line drawing requires that all the lines be black and sufficiently thick to avoid breaking up when the art is reduced to its final reproduction size. Toned images—anything in shades of gray, such as a black-and-white photograph—are screened, converting the tonal gra-

dations into black dots of varying sizes. The result, called a *halftone*, provides the illusion of continuous tones, complete with shades of gray.

For color originals, the art is separated into four separate plates, one for each of the four process print colors—black, magenta, cyan, and yellow. When printed together, the screened dots on the four plates combine to give the illusion of the continuous-tone color original.

BLACK LINE DRAWING

Black line drawing, also known as pen and ink, is a precise skill requiring a steady hand. Styles can vary from a simple line to elegant stipple and hatching techniques. Inkings can be done on paper or illustration board, though I prefer to use matte acetate or mylar film, which receive ink nicely and are dimensionally stable.

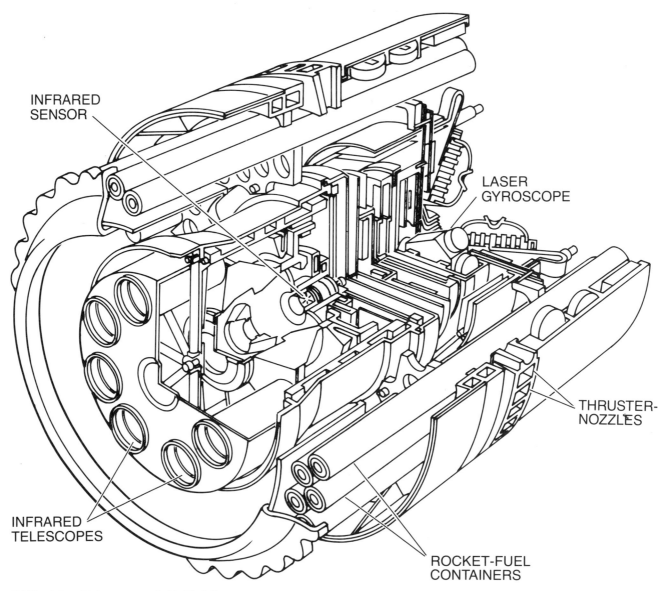

INFRARED SENSOR

LASER GYROSCOPE

THRUSTER-NOZZLES

INFRARED TELESCOPES

ROCKET-FUEL CONTAINERS

Orbiting infrared telescope executed in black line.

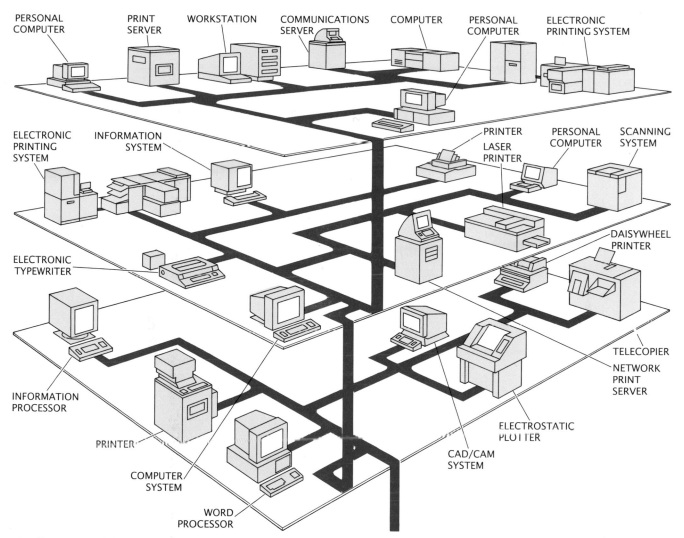

PERSONAL COMPUTER

PRINT SERVER

WORKSTATION

COMMUNICATIONS SERVER

COMPUTER

PERSONAL COMPUTER

ELECTRONIC PRINTING SYSTEM

ELECTRONIC PRINTING SYSTEM

INFORMATION SYSTEM

PRINTER

LASER PRINTER

PERSONAL COMPUTER

SCANNING SYSTEM

ELECTRONIC TYPEWRITER

DAISYWHEEL PRINTER

INFORMATION PROCESSOR

TELECOPIER

NETWORK PRINT SERVER

PRINTER

COMPUTER SYSTEM

WORD PROCESSOR

ELECTROSTATIC PLOTTER

CAD/CAM SYSTEM

Interoffice computer link-up. All pertinent features are easily conveyed with black line and one color.

BLACK LINE PLUS ONE COLOR OR GRAY

There are a number of ways to enhance a basic line drawing. Flat gray tones can be added by preparing an overlay in black indicating where the tone should print, with the tone's density specified as a black value of between 10 and 80 percent. The same procedure can be followed for the addition flat color tints. A little more skill and imagination is needed in planning this type of an illustration, but the additional color or gray tint goes a long way in dressing up a basic line image.

BLACK KEYLINE

In this format, a black line drawing is used as a guide for the photoengraver. Colors and tints are specified on a tissue overlay and the photoengraver strips in the appropriate screens for the various colors. This is a tricky procedure and requires an experienced artist with particularly good visualization skills, since the printer's proof will be the artist's first opportunity to see what the finished art will look like. Many times the results are disappointing, requiring costly corrections. This, combined with the somewhat dated effect created by tints and black lines, makes the keyline approach less than ideal for most jobs.

HALFTONE BLACK

The continuous-tone effect of a halftone lends itself to a great range of applications. The method is limited only by the artist's skill. Since the original art is reflective and camera-ready, the printing costs are only slightly greater than for line drawings, making this approach an attractive option.

This halftone black illustration was part of an article on optical illusions and the effects of light and shade.

HALFTONE COLOR

Halftone color plates are prepared in halftone black but are *printed* in a color specified by the artist, rather than in black. The original art must be painted in a darker tone than would normally be used for a halftone black print, since what is rendered as almost black on the art will reproduce in color as a medium tone in the final printing. Printing inks are lighter-valued than the original black tones used for original art, so if the halftone base art is prepared too light, the result will appear almost completely washed out.

Above is the actual black halftone rendering of a molecular structure; at left, the same rendering when printed in blue.

DUOTONE

The two-color approach of halftone black plus one color offers a very nice variation. In this technique, two engravings are made from a single halftone illustration, one for black and one for an artist-specified color. When they are printed together in register, the result is a very pleasing, deep, color-toned effect. For a blue–black duotone, for example, the halftone plate is clearly marked DUOTONE, BLUE-DOMINANT. This plate is given a longer exposure in order to make the color stronger. A second plate is printed in black and underexposed, so it will print with less intensity than the blue. The result is a rich, blue-dominant image.

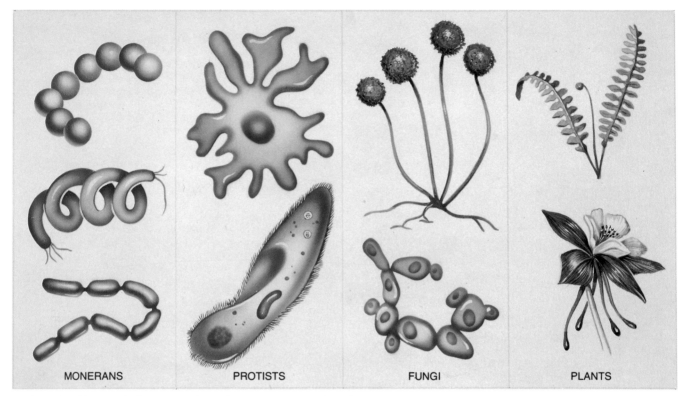

MONERANS   PROTISTS   FUNGI   PLANTS

Biological and botanical subjects printed as a duotone, to give the effect of a third color.

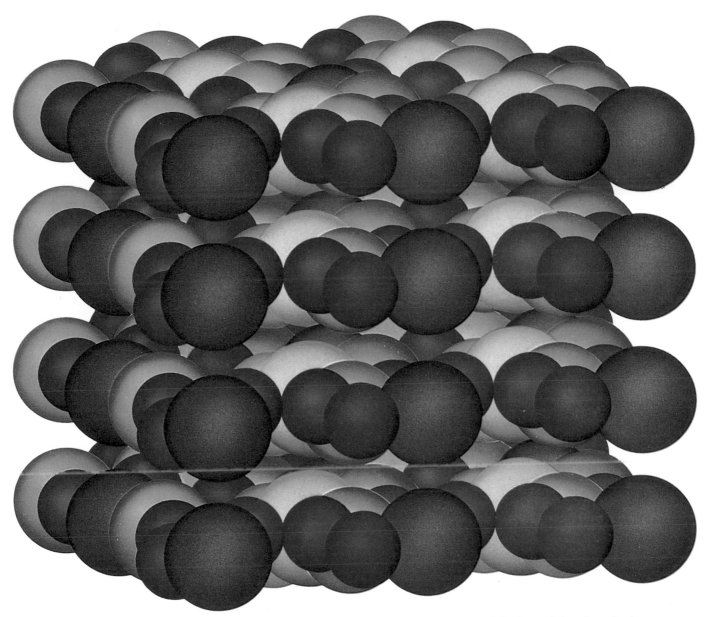

Full-color rendering of a molecular structure.

FULL COLOR

Nothing approaches the impact of a full-color illustration—it gives a skilled artist the fullest possible range of expression. Magazine illustrations are almost always in color; although full-color textbook illustrations were once rare, textbook publishers have more recently discovered what the magazine people learned long ago—the sales generated by the impact of full color more than compensate for the higher art and production costs. The resulting fierce competition to produce increasingly attractive publications has led to an increased demand for color illustration.

PREPARING FINAL ART FOR REPRODUCTION

I remember how disappointed I was early in my career when an important piece of art did not reproduce as I had expected. I have since learned how to prevent the more common reproduction problems. A little knowledge, skill, and planning can head off most of the problems that can occur as art goes through the various stages of engraving and printing.

In addition to the specific suggestions that follow, some general caveats pertain to all types of art. Each piece of art should be clearly marked for its necessary reduction percentage and its art form—line, color, halftone, duotone, or whatever applies; the same goes for all overlays. All tints should be clearly specified. Finally, protect your art by covering each illustration with an overlay of tracing paper or acetate. And above all, leave nothing to chance—the decisions left to someone else are the ones most likely to lead to disappointment.

SIZE

Although many illustrators routinely prepare their art at 300 to 400 percent of its final intended reproduction size, this approach has its drawbacks. The most obvious problem is that larger art takes longer to produce due to the simple reality of greater surface area to cover. Moreover, particularly large art boards do not fit on laser scanner drums, forcing the separator to prepare a film transparency of the art, an additional step that increases costs, decreases quality, and contributes to the loss of detail that inevitably occurs when large art is significantly reduced.

Therefore, I suggest that art be prepared at 150 to 200 percent of the final size. This allows enough detail to satisfy most situations, and also gives a good size ratio for selecting line weights that will not break up when reduced.

When the illustration is finished and ready for the printer, the final reduction size should be clearly indicated on the art.

CROP MARKS

Crop marks, which indicate the boundaries of an illustration, need to be accurately placed. There is nothing worse than finding an important detail has been inadvertently cropped out in the final reproduction, often in favor of a background area that could have been eliminated instead.

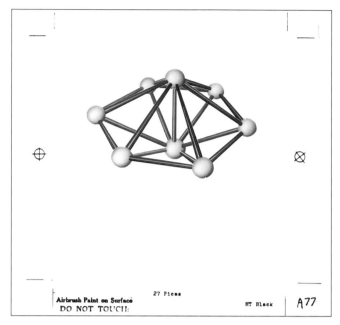

The base illustration on bristol board. This is the typical method of preparing an illustration for two-color reproduction. Note the register marks, crop marks, and labeled information.

REGISTER MARKS

When working with an overlay (such as for labels or tints), it is necessary to apply register marks on both the overlay *and* the art board, so that the printer will have a clear indication of exactly where the overlay should position. This also eliminates the possibility of the engraver accidentally flopping an overlay.

AIRBRUSH ART

Because airbrush art can be marred by even the slightest casual contact, art boards for airbrush art should include a warning sticker as follows: AIRBRUSH PAINT ON SURFACE! DO NOT TOUCH!

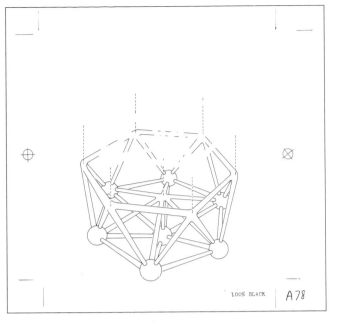

100% BLACK   A78

Overlay 1: A black ink drawing on mylar.

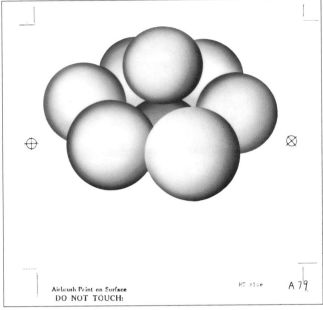

Airbrush Print on Surface
DO NOT TOUCH:

HT File   A79

Overlay 2: An airbrush rendering on mylar.

The final printed result.

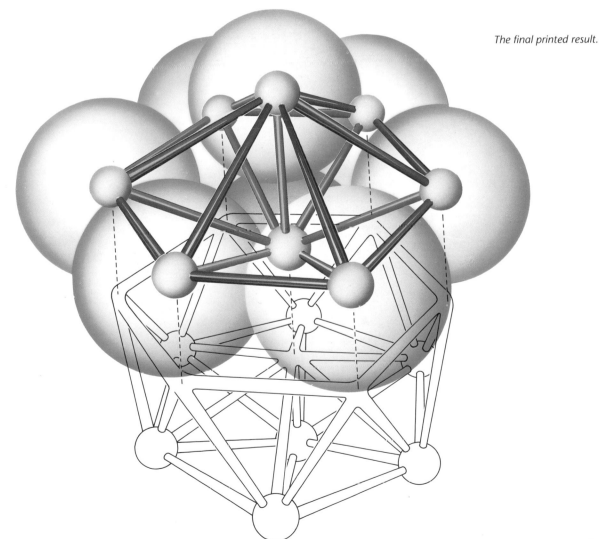

CHAPTER THREE

CLASSIC MODES OF PRESENTATION

While the style and sophistication of scientific illustration techniques have in some ways kept pace with the scientific advances they document, it is also fair to say that certain tried-and-true presentation approaches will almost certainly be the mainstays of any scientific illustrator's repertoire. The exploded view, the cutaway view, the flowchart—these illustrative techniques, along with the others discussed in this chapter, have proven their worth again and again over the years.

This is not to suggest that every project will lend itself to one of these presentation modes, nor to imply that even the most thorough knowledge of these techniques alone will constitute the comprehensive range of skills required of a successful scientific illustrator. To be sure, any good artist is constantly thinking about new approaches, innovative styles, and creative problem solving, and the very parameters of a given assignment may suggest a new solution technique.

Nonetheless, it is unlikely that even the most far-flung scientific or artistic advances will decrease the utility of the exploded view, and a great many projects will lend themselves to one of these classic approaches, rather than to something new. A solid grounding in these techniques is essential, and serves as a base from which to explore other possibilities.

Phantom view of plasma in the shape of a torus, confined in a magnetic bottle, with a cross-section view of the magnetic field configuration shown in black line.

Exploded Views

Exploded view drawings, most frequently employed to show mechanical assemblies, are used extensively in the manufacturing and service industries to train personnel in the assembly and servicing of modern machines. As the name implies, an exploded view shows all of an object's component parts removed from the main structure and spread out, generally in the reverse of the actual assembly sequence. The advantages of this approach become apparent with just a casual glance at such an illustration—even the most intricately designed device will yield its secrets when separated and stretched out in an exploded view. The consumer has become increasingly familiar with this technique through the advancing popularity of products that require assembly, such as low-cost furniture, barbecue grills, and so forth, the instructions for which usually depend on exploded views.

The choice to use an exploded view (or to use any of the other presentation modes presented in this chapter) is generally dictated more by what the artist is trying to show than by the subject matter. Through this technique, the artist enables the viewer to see and identify the characteristics of each separate piece of the subject.

Most exploded views are done as simple line drawings. Full-color renderings are sometimes used, however, usually in advertisements and brochures where the client wants to show a product's assembly features in the most attractive fashion.

Whenever possible, I disassemble the object to be illustrated in my studio, where I can either photograph the parts or draw them directly. The object is then reassembled in reverse order to check the accuracy of the drawing. Sometimes, however, this is not possible; in the case of the tank gauge and float shown here, for instance, the drawings were made from blueprints prior to the actual manufacture of the product.

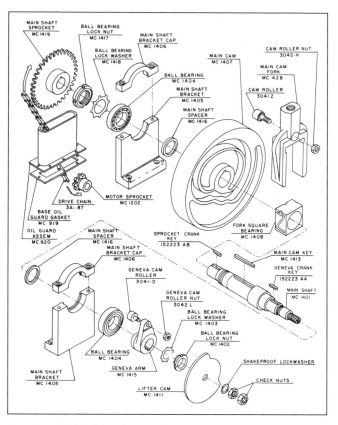

In order to fit this illustration on the page, it was necessary to "fold" the assembly. The dash-dot line represents the center line of the components. The trimetric drawing was developed from blueprints and executed in black line for a training manual.

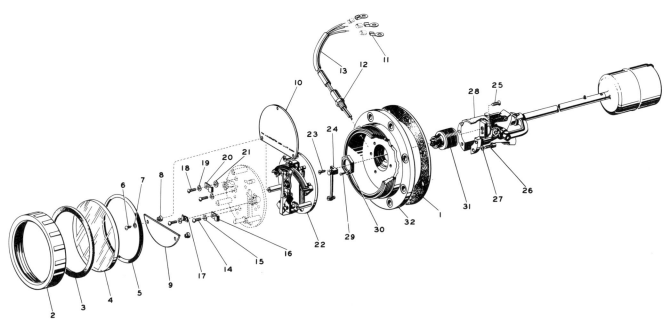

An exploded view of a tank gauge and float, executed as a line drawing. Parts are identified numerically and cross-referenced with a parts list. The outside retainer (2), gasket (3), glass (4), and O-ring seal (5) were overlapped to save space.

This example, while not an exploded view in the traditional sense, is based on similar principles. The parts remain engaged, rather than being separated, as is normally the case. The mechanism had to be taken out of its operational position in order to view the transfer of energy from the winding stem to the storage spring to the hands.

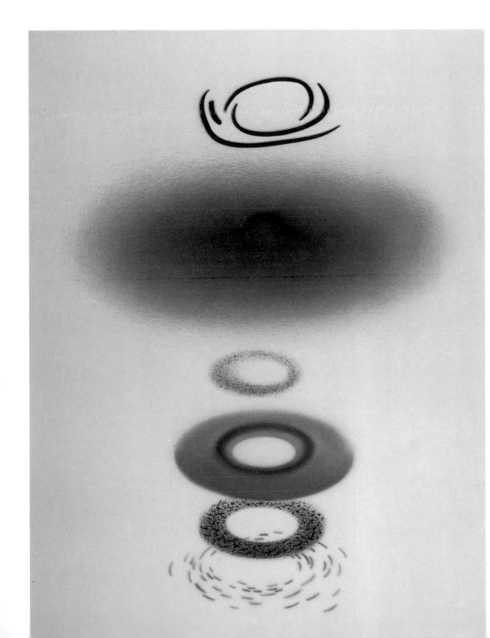

The Andromeda Galaxy is a complex phenomenon—it exists in a single plane with all its components sandwiched together. Pulling it apart in this unusual exploded view made it possible to identify the different sections. The drawing was done from sketches and notations supplied by the author regarding the relative size and proportions of the various components. Color was used for the further differentiation of elements.

A Practical Example: The Astrolabe Exploded

The astrolabe was the definitive scientific instrument of the Middle Ages, used for astronomical and terrestrial observations and as an all-purpose surveying tool before the advent of the sextant. It was also valued as an important computing device, determining the local time, the position of celestial bodies, and the direction of true north. As you can see from these drawings, it was quite an ingenious and sophisticated instrument, and re-creating it was no easy task.

The first step of the assignment involved a careful examination of the photographs—my only reference was a front-view photograph of the astrolabe and a few pictures of the interchangeable plates. Visualizing the relationship of the various parts drew upon my extensive experience working with mechanical objects.

The key was in the intricate network of circles. I realized that they had an interdependent geometry, but I had to figure out just what their interrelationship was before I could begin to transfer the information to the drawings. It was evident from the reference photograph that each succeeding circle had a slightly larger diameter. With the aid of a compass, I located the centers of half a dozen circles, suspecting that this would indicate a line along which the remaining centers would fall. This thought was soon confirmed: The individual circles' center points form a straight line heading toward twelve o'clock. Having determined this much, it was easy to locate the remaining centers. The points and measurements were then used to plot the circles for the exploded view drawing and the stereographic projection.

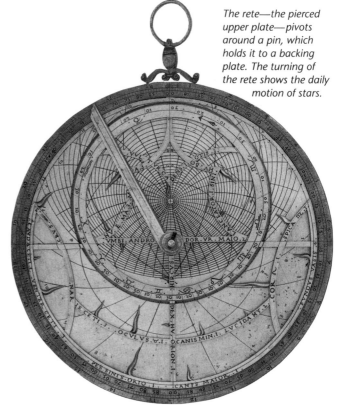

The rete—the pierced upper plate—pivots around a pin, which holds it to a backing plate. The turning of the rete shows the daily motion of stars.

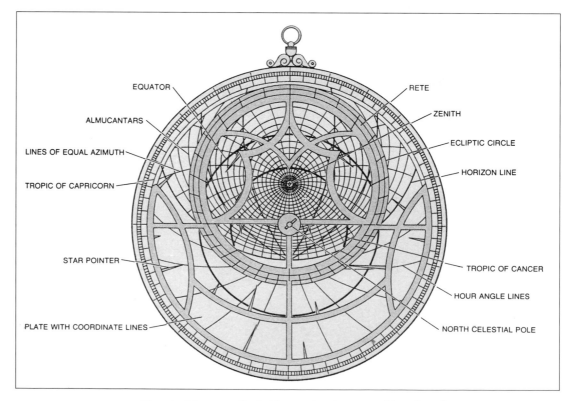

EQUATOR

ALMUCANTARS

LINES OF EQUAL AZIMUTH

TROPIC OF CAPRICORN

STAR POINTER

PLATE WITH COORDINATE LINES

RETE

ZENITH

ECLIPTIC CIRCLE

HORIZON LINE

TROPIC OF CANCER

HOUR ANGLE LINES

NORTH CELESTIAL POLE

The astrolabe was calibrated to record and survey a wide variety of geological and celestial movements.

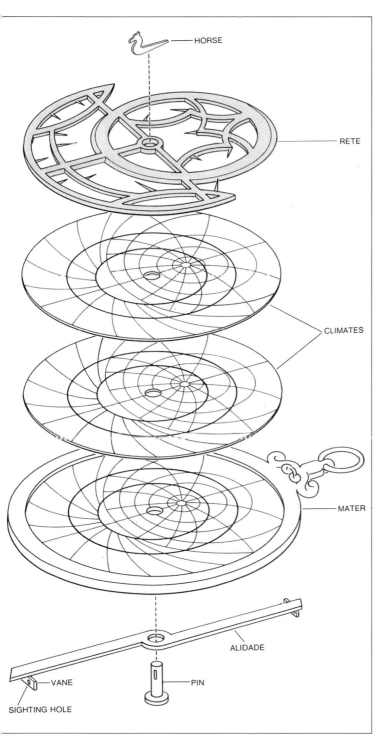

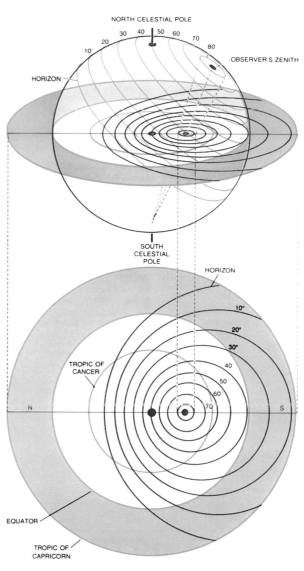

This stereographic projection of almucantars (circles on the celestial sphere running parallel to the horizon) drawn on the astrolabe shows that circles on a sphere remain circles when projected onto a flat plane; they do not, however, have a common center.

Exploded view of an astrolabe: The mater is the main body. The climates, or plates, are engraved with coordinate lines for different latitudes. The alilade, used for sighting celestial objects, rotates freely at the back of the unit. The pin slides through the center of all the plates and is held in position by the horse's head.

CUTAWAY VIEWS

The cutaway view, unlike the exploded view, shows all of an object's parts in their assembled and operating positions, thereby allowing the relationship and interaction of the components to be easily followed.

In this type of illustration, part or sometimes half of the structure is "cut away," creating a window that reveals the inner workings. The technique requires skill in trimetric projection (see Chapter 6) and perspective drawing. Moreover, convinc-

ing and accurate cutaway views necessitate an exact knowledge of the form, substance, and structure of the subject, particularly for work in the medical and biological sciences.

Artists attempting to draw cutaway views need superb drawing skills, as well as the ability to visualize internal and external structures in their correct relationships. For complex medical and biological cutaways, I usually consult several textbooks and extract whatever information I need from each one.

The outer portion of the eye is cut away in this illustration, prepared to show a surgical procedure performed with laser beam.

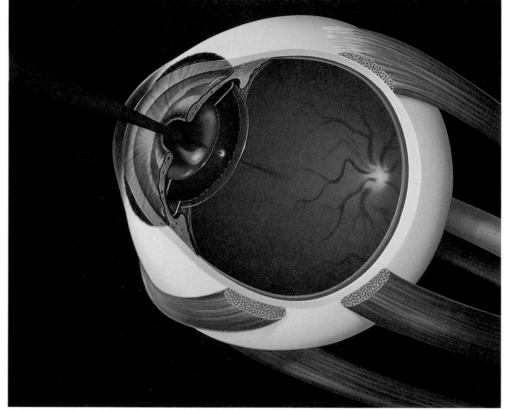

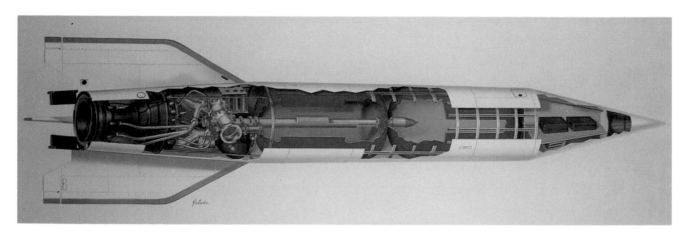

For this cutaway rendering of a German V-2 rocket from World War II, color coding is employed to indicate fuel tanks, the engine, and the payload.

By using the cutaway view to show these two very different vessels, the ancient Golden Hind (left) and the modern research submarine Trieste (below), the integrity of the whole is maintained while interesting comparative details in the inner space of each can be explored.

CROSS-SECTION VIEWS

With this approach, the artist "slices" an object in half to reveal its inner components and structure. By targeting the slice at specific internal components, the artist can show the viewer both the placement of internal structures and the detailed composition of individual components.

Careful planning is required in selecting the angle and plane that will show the viewing area to best advantage. Objects can be drawn three-dimensionally or as flat head-on views.

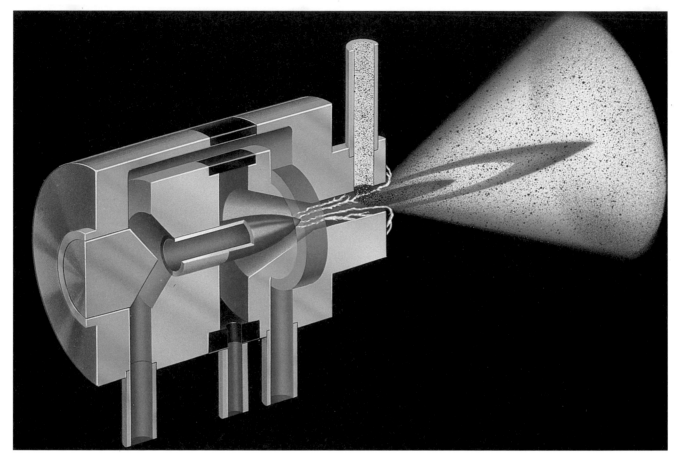

Above: Cross section of a plasma gun used for spraying an anti-corrosion coating. This viewing angle makes it easy to identify the specific part of the gun being examined and exposes the channels through which the coating material flows and is mixed.

Right: This cross section shows surface features, the mid-Atlantic ridge, and rising plumes of molten magma.

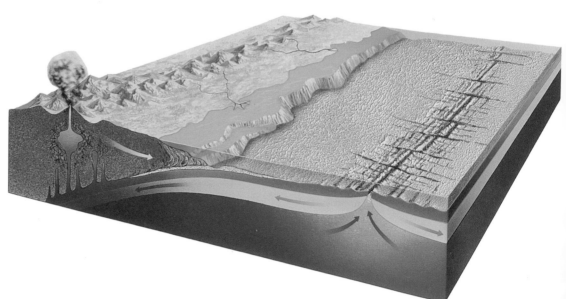

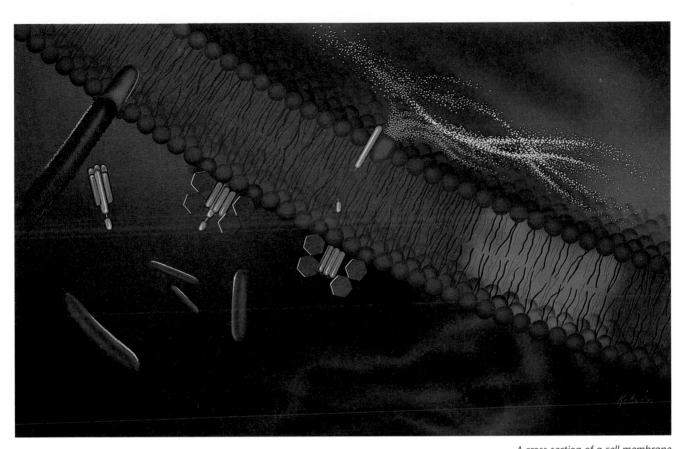

A cross section of a cell membrane showing organisms passing through from one side to the other.

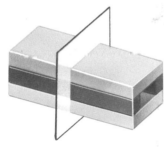

On the left, one object is cut in three different planes; on the right, the resulting cross sections.

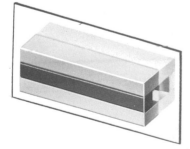

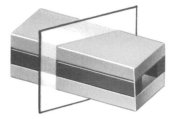

PHANTOM VIEWS

The phantom view is among the more elegant and versatile of the specialized illustration styles. With this method, solid structures become translucent, making it possible to see components hidden in inner spaces without disturbing the spatial relationship between the visible inner space and its surrounding environment. The technique works well with both mechanical and biological subjects, and can be used in combination with cutaways or cross-section views.

The ghosting effect of the phantom view allows the subject's outer structural integrity to remain intact, making this technique particularly desirable when there is no need to show anything involving sections of the outer structure. For instance, if I want to show the human heart in the context of the body, I simply ghost the outer body and focus on the heart. If I had to show the skin, fat, muscle, and bone as well, I would choose a combination of cutaway and cross-section views.

Inner supporting girders and guide wires can be seen in this ghosted view of a dirigible's helium-filled outer envelope.

The steel armor plate of this M-1 tank is partially ghosted to show the crew members at their stations within the vehicle.

The ghosted camera frame permits an interior view of this electronic video still camera.

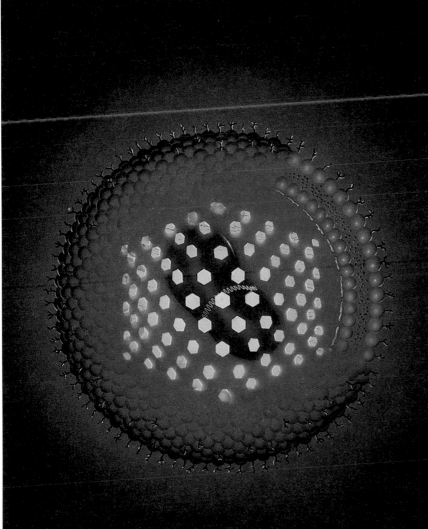

The protein outer shell of a herpes virus is ghosted here to show the organism's core structure.

Generic Illustrations

Articles that present a point of view or enter into a general discussion about a type of product may be treading on dangerous ground if the text or supporting illustrations unwittingly implicate or favor a particular manufacturer. In an attempt to inform the public on fax machines, for instance, it would be prudent to show a generic machine that incorporates the main features of many fax machines. This usually entails disguising the product to avoid making its brand readily identifiable.

Fortunately, there is no need to show all the features of *every* model of a given product, since the depiction is meant to be generic and not necessarily a composite representation. Sometimes, however, a product includes a feature that can appear in two or more mutually exclusive yet equally impor-

tant configurations. Some electronics firms' multiple-disk CD players, for instance, feature a rotating carousel tray, while other firms use a "stacker" cartridge mechanism for the disks. In such cases, this might necessitate two drawings to adequately represent both types.

With this in mind, these line drawings have been simplified to show the evolution of farm machinery, beginning with the earliest steam tractor and proceeding right up to today's modern diesel-powered versions. The tractors underwent some cosmetic redesigning to avoid identifying the ones that were used as the models. Altering the pattern of the wheel tread, restyling the cab seats, and changing the chassis configurations resulted in an authentic-looking, generic farm machine.

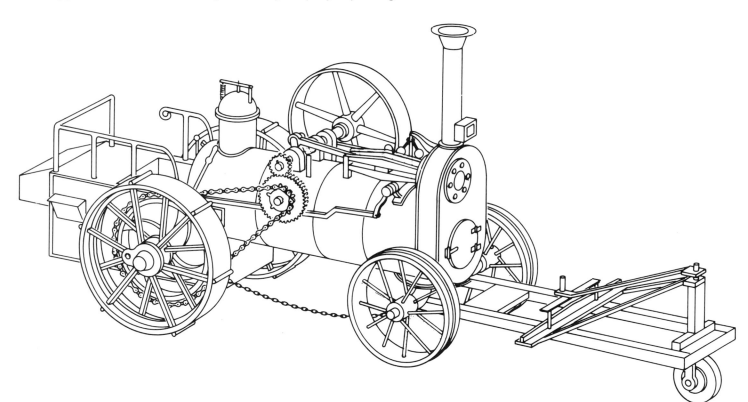

Early steam-driven farm tractor, c. 1860.

By the 1930s, tractors had incorporated a more sophisticated engine design.

This four-wheel-drive modern tractor generates 300 horsepower and is articulated at a center hinge, allowing the rear wheels to follow the same path as those in front.

FLOWCHARTS AND DIAGRAMS

Flowcharts and diagrams provide the artist with a creative means to display the sequence of events of a given process, tying the various components together in one comprehensive graphic. Here a wide variety of design and styling options are possible, ranging from simple line drawing to full color. Whatever stylistic route is chosen, however, the essential aspect of this type of chart is its readability—attention must be paid to careful organization of information, and complicated pathways and lines crossing over each other should be avoided. With this goal in mind, it is the artist's responsibility to suggest alternative ways in which charts can be planned, so as to clarify without eliminating any information along the way. Contrasting color tints help to differentiate component groupings, while stylized graphic symbols can simplify system identification and lend an attractive touch.

Flowcharts and diagrams fall into a number of types and styles, several of which are shown here. In choosing which type to use, the artist must employ and adapt the form best suited to the illustration's purpose.

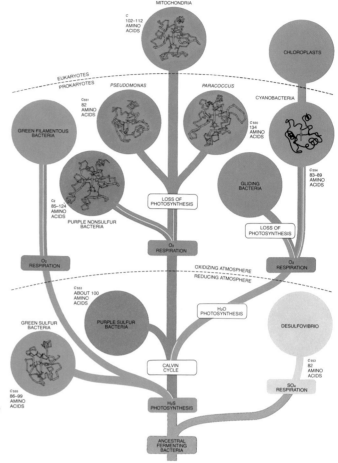

In this "tree"-type diagram tracing the evolution of bacterial photosynthesis and respiration, color is used to differentiate between separate information groupings. Close examination reveals that this rendering, prepared for a professional audience, is a surprisingly easy-to-follow chart of extremely sophisticated and technical information.

Drop shadows, such as those used here, can add dimension to an otherwise flat-looking chart. With the acetate mask for the background offset from its original position, a darker value of the background color was blown in with the airbrush. The angle of the shadows was an artistic determination.

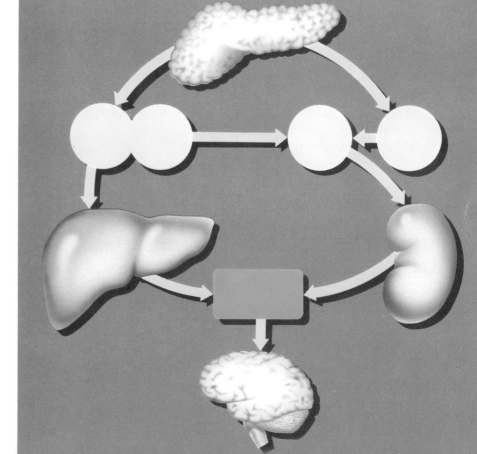

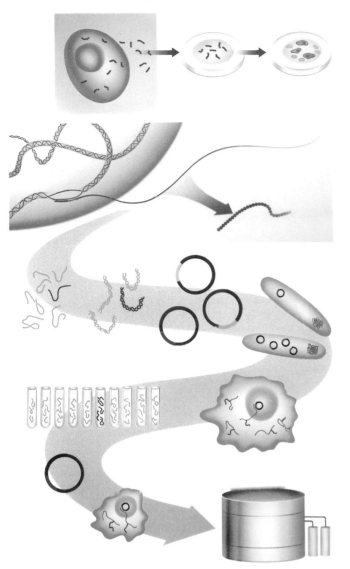

Fitting an illustration within an allotted space sometimes calls for a little ingenuity. Here a folded arrow designates the pathway taken in a genetic cloning process. Had there been more room, the arrow would have been shown fully extended and straight.

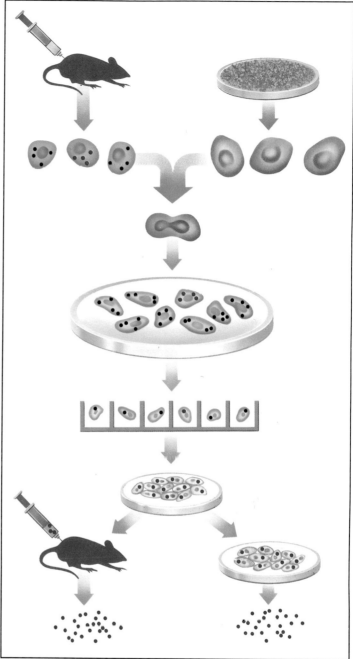

In this simplified version of a laboratory experiment, a rat is injected with a cell culture, while a petri dish contains the same culture. The chart shows the stages of the experiment, culminating in two variations of antibodies.

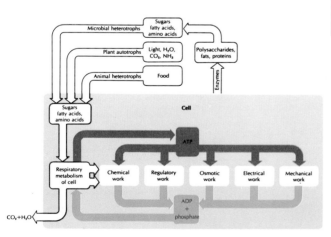

A simple flow diagram designed with arrows and boxes.

CHARTS AND GRAPHS

Traditional preparation of charts and graphs calls for precision inking skills and a reasonable grasp of mathematics. Every point of numerical datum has to be individually plotted on the chart's vertical and horizontal axes. Today, however, computers have replaced much of the work previously done by hand, and the necessary information can be fed into a computer and plotted quickly and easily. Computer graphics systems offer a full range of design and color possibilities as well.

But the computer, like any other tool, is only as good as the person using it. Fundamental to scientific illustration is the capability to interpret raw data and make it work artistically, regardless of the method. Furthermore, there are limitations to any computer program, which is particularly noteworthy in light of the fact that the graphic computer systems priced to fit most art studios' budgets cannot match the color quality and subtle effects possible with the very sophisticated, more costly equipment.

In terms of scientific illustration, the most valuable application for a reasonably priced computer system is in graphs, charts, icons, generic symbols, simple graphics, and, most importantly, generating the reference data needed to produce complex geometrical drawings, as seen in some of these diagrams. For this type of work, mathematical formulae are converted into numerical data with the aid of a hand-held computer and used to plot the points on a simple X–Y graph. Once all the points are plotted, I connect them with a smooth line, making slight eyeball modifications as I go along, resulting in the sort of version shown at the middle of page 59. For the more complex version at the bottom of page 59, well over a thousand points had to be plotted and connected.

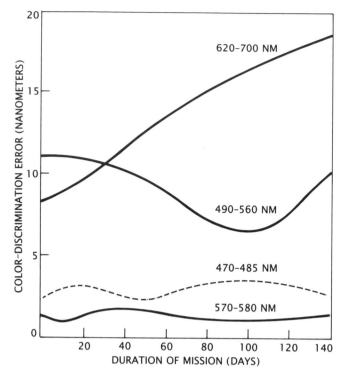

A simple line graph, plotting an astronaut's ability to identify color accurately over the duration of a space mission.

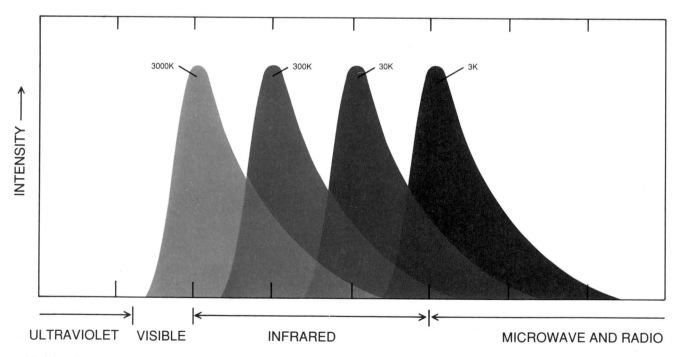

A black-line chart with full color, showing the electromagnetic spectrum.

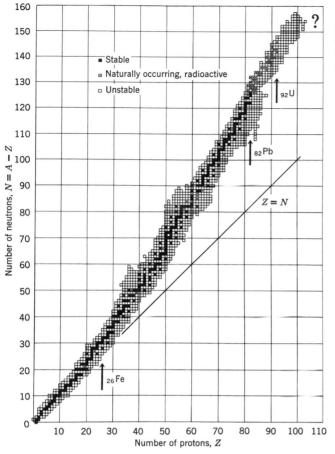

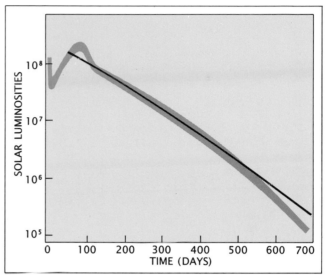

Grid lines were added to this graph of protons and neutrons in order to show a more precise location of the points, which are either stable, unstable, or in a radioactive condition.

Above: In this example, a line graph showing the duration of solar luminosities is supplemented with color tints. Because the two curves intersected at several points, color was used to distinguish one from the other.

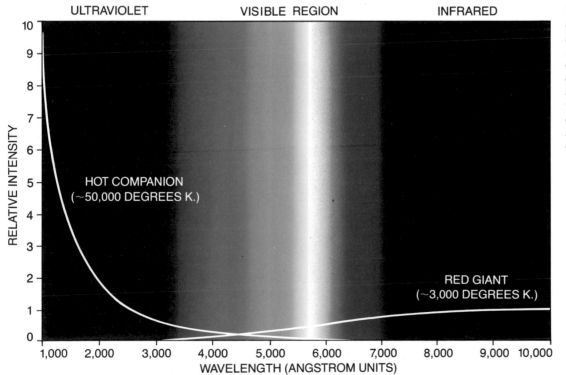

White curves plotted against a full-color background depict the relative intensity of a red giant star in comparison with its hot companion. This was executed in airbrush to obtain a soft blending of colors.

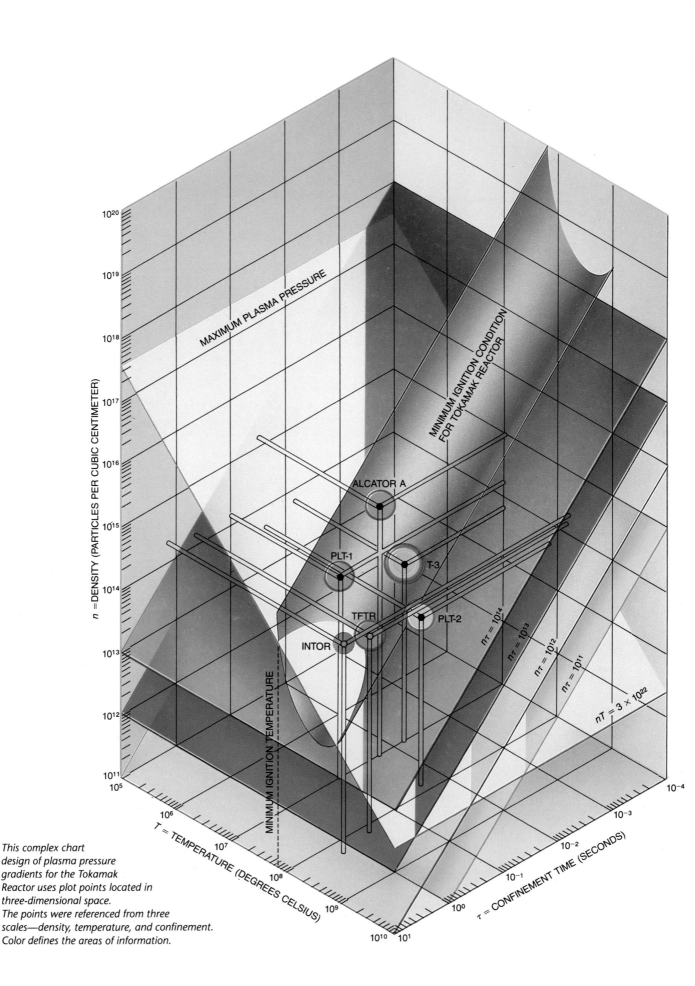

This complex chart
design of plasma pressure
gradients for the Tokamak
Reactor uses plot points located in
three-dimensional space.
The points were referenced from three
scales—density, temperature, and confinement.
Color defines the areas of information.

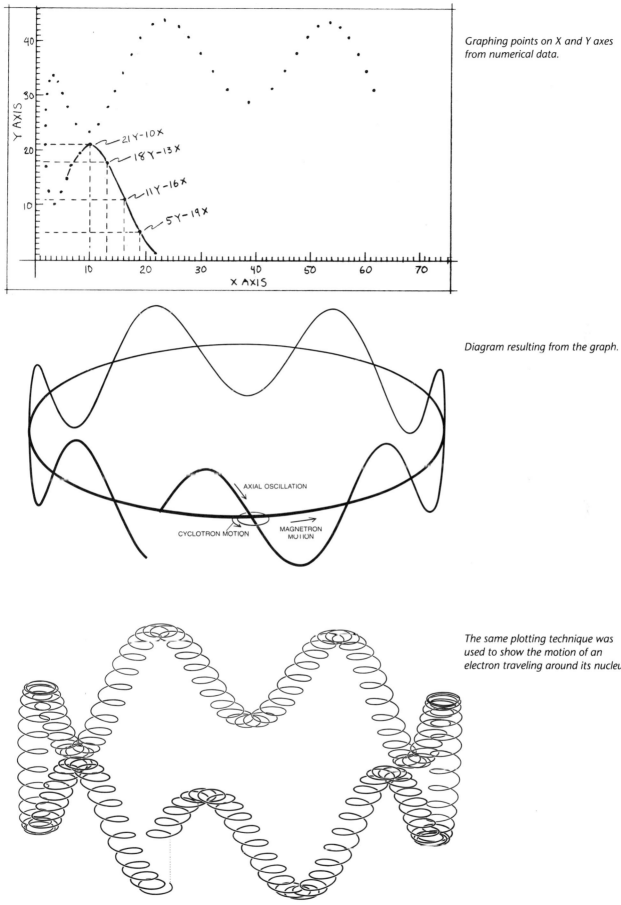

Graphing points on X and Y axes from numerical data.

21 Y – 10 X
18 Y – 13 X
11 Y – 16 X
5 Y – 19 X

Diagram resulting from the graph.

AXIAL OSCILLATION

CYCLOTRON MOTION

MAGNETRON MOTION

The same plotting technique was used to show the motion of an electron traveling around its nucleus.

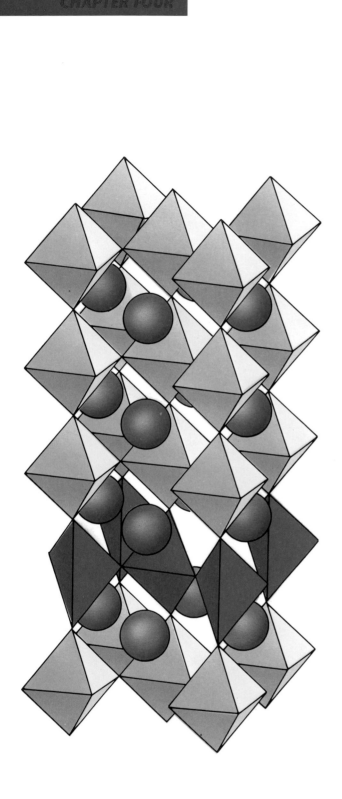

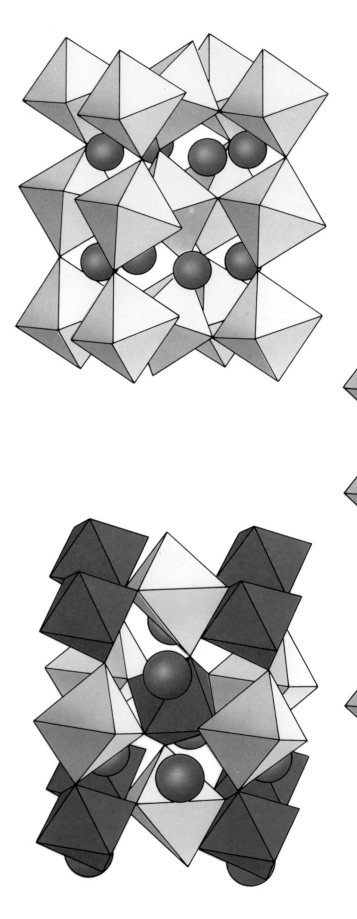

CHAPTER FOUR

WORKING WITH MODELS AND PHOTOGRAPHY

When drawing structures with complicated geometries, the accurate representation of angles and proportions is of extreme importance. In these instances, it is particularly helpful to build a model. No matter how developed an artist's visualization and execution skills may be, there is no substitute for holding a three-dimensional model and examining all the possible approaches for showing the structure to its best advantage.

The job of the artist is to alter and adjust the various elements of the illustration so that they will look "right." The model serves as a guide to that end. It can be rotated and examined from every conceivable angle, saving hours and possibly days of drawing time.

In addition, the model can be photographed from various perspectives, so there is a definite advantage to having a basic understanding of cameras and photographic techniques. The ability to take high-quality reference photographs helps to solve drawing and perspective problems that might otherwise be difficult as well as time-consuming.

A complex model incorporating octahedrons, polyhedrons, and tetrahedrons was constructed and photographed to produce these structures. Geometric shapes are the basis for many scientific illustrations, so model-making skills are invaluable.

CONSTRUCTING MODELS

I can usually tell from the very outset of a project if a model will be needed. To a large degree, this ability comes from experience. While it is always possible to draw directly from the model itself, it is faster and more accurate to work from the subsequent photograph.

I maintain a small permanent collection of human anatomical models, but all the other models I use must be built from scratch. Most of these are not very demanding in terms of materials or equipment. As an inveterate collector of all kinds of odds and ends, I am prepared for most projects. The tools and materials I use most frequently are listed here.

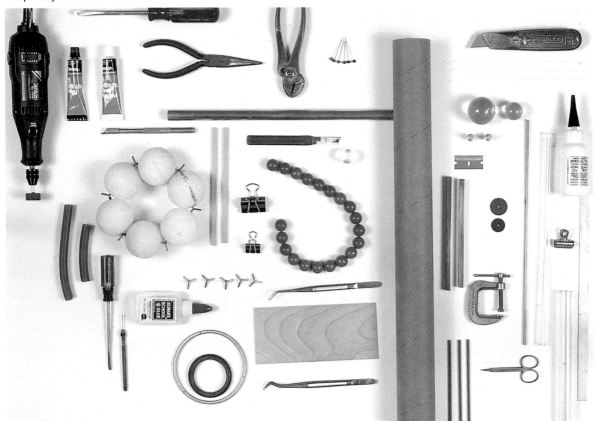

Tools	Rasps	**Materials**	Brass rods, tubes, and sheets	Cardboard tubes
Pliers	Clamps	Duco cement	Solder	Cardboard construction paper
Screwdrivers (Phillips-head, standard, and miniature)	Weights	Epoxy glue	Styrofoam	Foam board
	Tweezers	Spray adhesive	Magnetic marbles	Cord
Drills	Sandpaper	Tape	Cork balls	Rings
Clamps	Dremel moto tool	Balsa wood	Springs	Rubber tubing
Hemostats	Pins	Bass wood	Molecular construction set	Magnets
Dental pick	Razor	Acetate sheets		Beads
Files	Hot-glue gun	Plastic rods and tubes	Modeling clay	masking tape
Burrs		Lucite rods and tubes	Wire	

EXAMPLES OF MODELS

A DNA HELIX

The heavy research now taking place in the field of genetics has caused the double helix structure (commonly called DNA) found in molecular cell biology to be a recurring subject of scientific illustrations. Drawing a double helix (or a single helix, for that matter) presents problems in perspective that can be easily solved with a simple wire model construction.

Start with either a soft copper or zinc wire, readily available in hardware stores. Wind it carefully around a pencil or, if a larger coil is desired, use a brass or cardboard tube. The result should look like a spring.

Next, slip it off the pencil or tube. Holding it at each end, stretch it to about twice its coiled length—it should form a perfect helix. The length to which the wire is stretched determines the helix's angle.

There are other possibilities as well. Brass strips wound in the same manner around a rigid tube, for example, will form a helix that can be bent into almost any shape without losing its helical twist. Another method calls for wrapping masking tape in a spiral pattern around rubber tubing and inserting a heavy copper wire into the tube, allowing the helix to be bent into a variety of configurations.

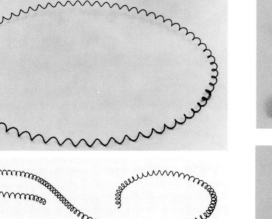

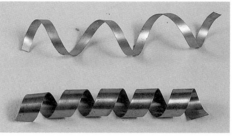

Brass strips wound around a rigid tube can also be used.

Yet another method employs masking tape wrapped around rubber tubing with a copper-wire insert.

Copper wire, once wrapped around a pencil and stretched, yields an ideal helix form.

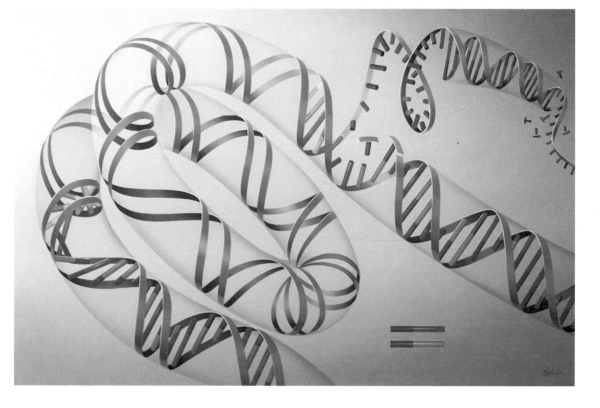

Schematic representation of coiled DNA.

MOLECULAR MODEL GEOMETRY

Illustrating for science frequently involves rendering geometric forms, many of which are based on the square, circle, triangle, pentagon, and hexagon. The tetrahedron, for example, which is commonly associated with chemistry diagrams, has four equilateral triangular faces. Also common is the icosahedron, which consists of 20 equilateral triangles. All of these shapes lend themselves nicely to model formations.

Stiff construction paper is an easy-to-use material that is particularly well-suited to geometric applications. In addition, kits for constructing geometric molecular models can be purchased through catalogs from scientific supply houses. The kits contain joining devices with various angles and plastic connecting rods. They are easy to use and handy to have.

For constructing three-dimensional skeletal models, my favorite material is balsa wood sticks. The sticks are easy to cut and shape and surprisingly strong when assembled.

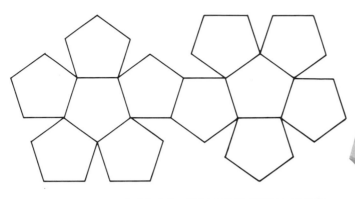

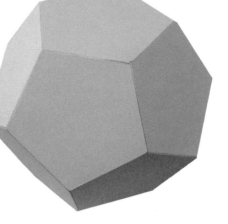

As this dodecahedron demonstrates, once the geometric layout is drawn (above), the cutting, folding, and gluing are simple (right).

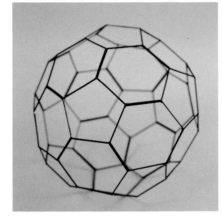
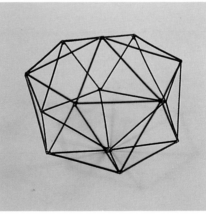
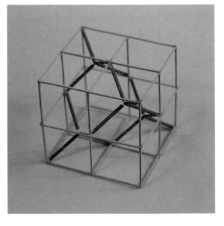

Balsa wood stick models of molecular structures.

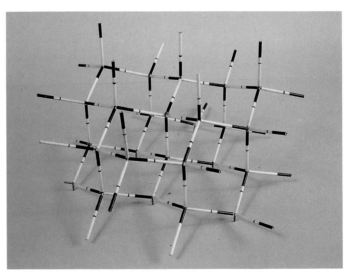

Molecular model constructed from a kit.

BEAT WAVE

For a *Scientific American* article on accelerated plasma particles, I needed to show longitudinal waves moving through electrically charged plasma, a phenomenon known as a beat wave.

The raw data was supplied as a computer-generated side-view drawing showing the configuration and spacing of the waves. Measurements taken from this drawing became the basis for the balsa wood and paper model shown below, which incorporates all the necessary information except for one important element: The waves are circular in a frontal view but ellipsoidal when seen in cross section. Rather than complicating the model, I adjusted the pencil sketch later on, which is why the waves appear thicker in the illustration than in the model.

The wooden dowel served to identify the exact wave centers and became the midline guide for the plasma environment shown in the final illustration. The final image was modified to reflect more current data, accounting for the differences between the model and the art.

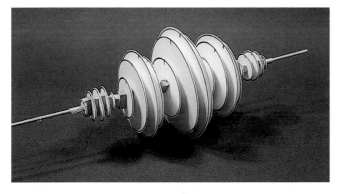

The model for the beat wave illustration was constructed from three-ply paper, balsa wood spacers, and a quarter-inch (6-millimeter) dowel. The components' size and spacing were derived from a side-view computer drawing.

Interference pattern of beat waves (yellow) impacting undisturbed plasma (red).

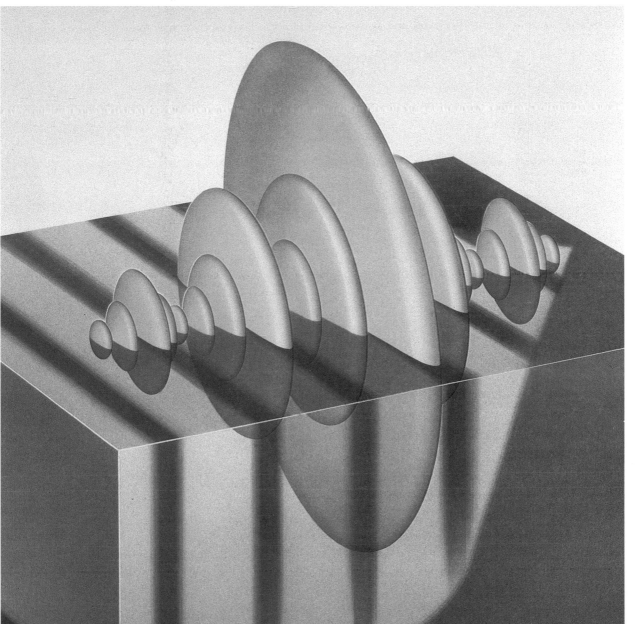

REFERENCE PHOTOGRAPHS

Drawing proper angle and line configurations directly from a complex model requires a series of trial-and-error sketches and necessitates taking exact measurements, adjusting them for scale, and plotting them on the drawing. By photographing the model, I accomplish all of this in a matter of minutes and at the same time acquire ultimate control over my reference material and expand my options.

The camera that best suits my studio needs is the Polaroid 600SE, a versatile and practical tool that serves as the backbone of my studio. It has a Mimaya interchangeable lens with a very wide aperture range, from f/64 to f/4.7. Used in conjunction with my enlarging and reducing copier and darkroom enlarger, it is effective for solving a variety of illustration problems.

Unfortunately, the 600SE cannot take focused pictures from closer than 3.5 feet (1 meter). And since my work requires a close-up capability of 3 to 4 inches (7.6 to 10.2 centimeters), it was necessary to modify the camera. The camera is equipped with a rear film pack feature that allows the film to be changed in the middle of a roll as the photographer's needs change. I used this special feature to design a removable viewing box, which I adapted to the back of the camera. The box has a magnifying-glass lens and a translucent screen that fits into the exact position the film normally occupies, thereby giving me a direct view of the subject to be photographed. Once I have set up the shot as I want it and have employed the appropriate close-up lens, I remove the viewing box, replace it with the film pack, and proceed with the first picture.

The Polaroid films I have found most versatile are Professional 667 black-and-white (ISO 3000) and Professional 665 black-and-white (ISO 80). The extremely fast 667 film allows me to use the camera's near-pinhole f/64 aperture setting to produce photographs critically sharp from foreground to background. This is particularly useful for pictures of small-scale model constructions that need to be photographed from within a few inches (5 to 13 centimeters) in order to produce images large enough to draw from. Even though these close-up photos often must be enlarged by as much as 10 times on the copy machine, the sharpness provided by the small aperture and high-speed film allows all this to take place without significant distortion.

The 665 film is much slower but has the advantage of supplying both a positive image and an extremely fine-grained film negative, all in one shot. I use this film when I need to enlarge an image to between 10 and 20 times its original size.

Working with a set of three close-up lenses in strengths of 1×, 2×, and 4× allows a series of lens combinations with a power range of from 1× to 7×. It takes a little experimenting with different lens combinations to get exactly the right setting for a given subject. In general, the distance from lens to model determines which lens combination to use—the closer the camera is to the model, the greater the magnification.

Above: The Polaroid 600SE, a particularly versatile camera, is ideal for photographing models.

Right: Assorted tools and materials for photographing reference models: view box, mounted tweezers, tripod, film pack, shutter-release cable, light meter, and three close-up lenses.

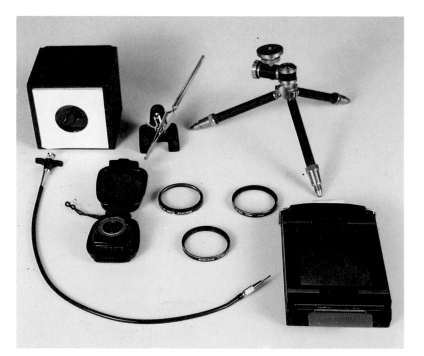

Front of viewing box, showing eyepiece lens.

Rear view, showing acetate viewing screen.

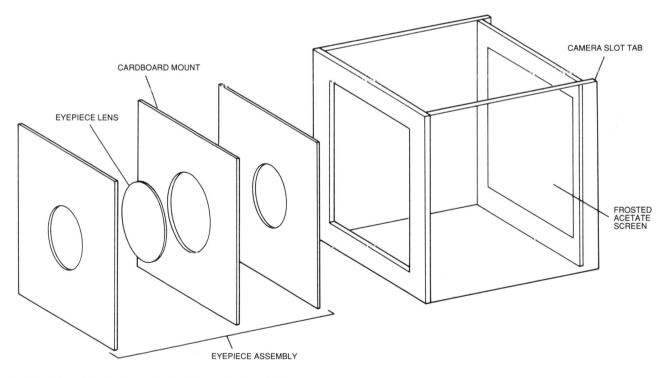

CARDBOARD MOUNT

EYEPIECE LENS

CAMERA SLOT TAB

FROSTED ACETATE SCREEN

EYEPIECE ASSEMBLY

Exploded view of viewing box, showing full construction details.

Photographing Constructed Models

Models to be photographed are positioned on a piece of construction paper, selected to give the best photographic contrast. Since I shoot with black-and-white film, the paper is generally black, white, or a shade of gray. Next, I decide what angle will best show the model's features and make certain that all the important elements will be in clear view.

Once the appropriate angle has been determined, the camera is mounted on a tripod and positioned 4 to 12 inches (10 to 30 centimeters) from the model, with a shutter-release cable attached to the camera. Keeping the safety slide in position, the rear film pack, loaded with 667 film, is removed from the camera and replaced by the viewing box. Using the camera's bulb setting and widest aperture setting (f/4.5), the shutter-release cable is engaged to hold the shutter open. This setup allows me to see exactly what will be exposed to the film.

Because the camera lens is positioned just a few inches from the model, everything appears out of focus when I look through the viewing box. This problem is solved by inserting a close-up lens on the front lens mount, which brings the subject into soft focus but is sufficient for composing the picture and making sure all the necessary elements are in view. Once I am satisfied, I lock the camera into position, remove the viewing box, and deactivate the cable release to close the shutter. Then I put the film pack back on the camera, remove the safety slide, and readjust the lens aperture to f/64—barely a pinhole, but enough to produce the extreme depth of field required for the shot. The part of the field that appeared slightly out of focus when I looked through the view box earlier is now razor sharp.

Because the film has such a fast speed, my desk fluorescent lamps can be used as the main light source. A quick reading of my light meter indicates a one-fifteenth-second exposure at f/64. After setting the exposure and aperture controls accordingly, I press the cable release, and 30 seconds later I have my first print, sharp and clear from foreground to background.

If the shot is unsatisfactory or if I just want to see how another view might look, I can simply leave the camera where it is and rotate the model in place, taking care not to change the distance from lens to model, and take additional pictures.

This procedure may sound more complicated than it really is. The time from setting up the model to finishing 8 or 10 photos usually takes about 30 minutes—time well spent, considering the time that might have been spent on false starts and wasted labor had I attempted to draw the model without photographing it first.

A tripod is essential for effective reference photography.

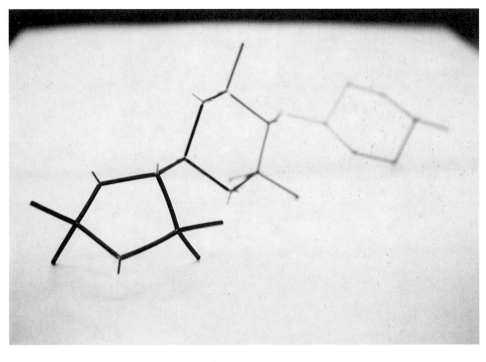

In order for enough light to enter the viewing box, the lens must be kept fully opened, causing part of the image to be out of focus. Here is how it appears through the viewing box.

Right: When the aperture is reduced to f/64—essentially a pinhole—the entire image comes into sharp focus. The final photo is seen here.

ON-SITE PHOTOGRAPHY

STEREO PHOTOGRAPHY

Not every situation lends itself to conventional photography. Large structures and equipment, for instance, cannot be brought into the studio and photographed. However, when alternative reference material is not adequate to draw from, the artist-photographer may want to visit the site of the subject and take additional photos.

Pictures for viewing in stereo are among the on-site photography methods that can prove very useful for special types of assignments. Unlike a conventional photo, which is a flat, two-dimensional image, the stereo image appears three-dimensional, making it possible to judge the spatial relationships between the various structures in the picture.

For stereo photography, the camera is set on a tripod and two shots of a section of the structure are taken, with the tripod moved laterally about 3 inches (8 centimeters) between shots. This is repeated until a pair of stereo shots of each section of the structure has been taken. When the paired photos are placed in a stereo viewer, the subject appears three-dimensional. The stereo photos, used in conjunction with some standard reference shots, are generally enough to draw from.

Mirror assembly for stereo viewer.

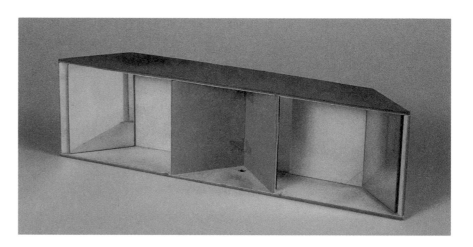

Mirror assembly in stereo box.

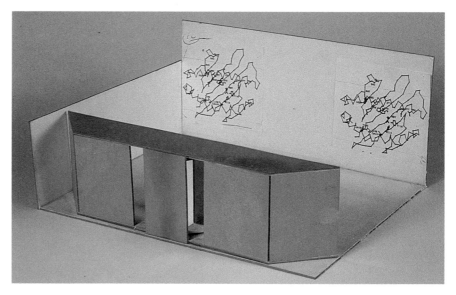

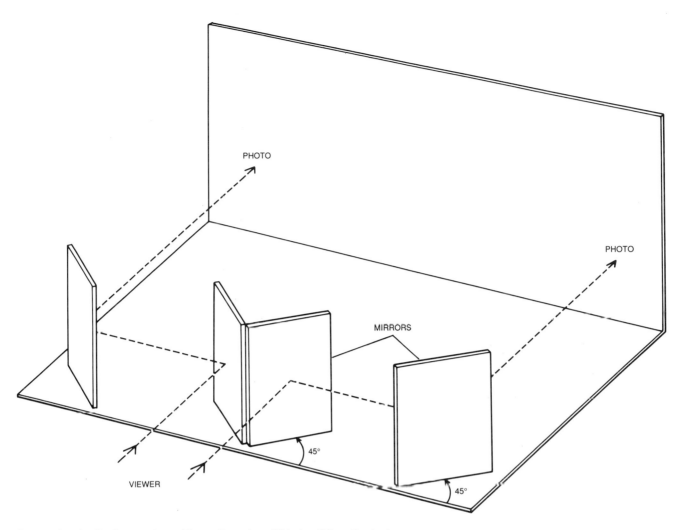

Construction details of stereo viewer. Viewer dimensions: 13 inches (33 centimeters) wide by 11 inches (28 centimeters) deep by 6 inches (15 centimeters) high.

PANORAMIC PHOTOGRAPHY

Panoramic photography is another on-site technique that can be utilized to photograph very large equipment or machinery. A subject 40 feet (12 meters) wide, for instance, obviously cannot be photographed in one shot. For this type of undertaking, I set my camera on a tripod and take a panoramic picture.

By maintaining the camera's distance from the subject and moving the camera and tripod laterally along a straight line in steady increments equal to the field of view, I can take a series of photographs along the length of the structure. Joining these pictures together results in a single image, with only minor adjustments needed to match up the segments.

PHOTOGRAPHING FOR EXPLODED VIEW ILLUSTRATIONS

For presenting several components of complicated machinery (gears, cams, brackets, and so on), an exploded view is the most effective way to show all the parts in their order of assembly. Drawing exploded views without vanishing points, and with all the parts shown from the same angle, is usually accomplished by using trimetric projection, with predetermined angles and ellipses. But photographing the parts first can save a great deal of drawing time and can be used instead of or in conjunction with the projection method.

Disassembling the object, arranging its various parts in their sequential assembly positions, and then photographing them that way would introduce perspective into the angles and ellipses, causing the parts to appear not to fit together as a unit. Instead, begin by constructing a simple jig or holding device to hold each item in position at exactly the same angle relative to the camera. After photographing each item in the jig, lay out the pictures in their respective positions.

Once the photos have been properly assembled, there is the option of doing a line drawing tracing, executing a black-and-white or full-color rendering, or using the actual photographs to show the exploded view.

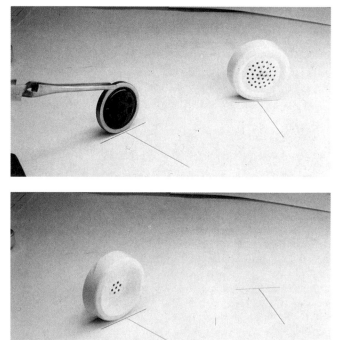

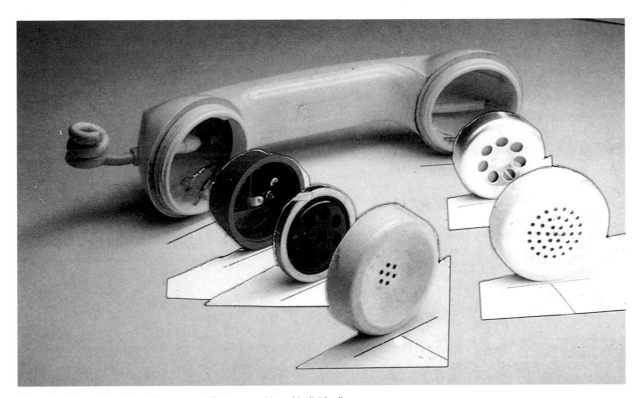

Photographic exploded view: Components first are positioned individually on alignment marks (left); the individual photos are then cut apart and positioned on the alignment axis to form a composite exploded view (above).

PHOTOGRAPHING FLOOR PLANS FOR PERSPECTIVE DRAWINGS

Architectural floor plans are commonly converted to perspective drawings, often for use in brochures promoting the sale or rental of available space. The problem of transforming a flat view into a perspective view presents another opportunity to save time through photography. The camera automatically records the images in perspective, making it a simple matter to locate the necessary vanishing points for the conversion.

When dealing with a single-floor layout, the photographs are taken directly from the plans; for a more complex layout, a tracing paper overlay is placed over the blueprints in order to trace the salient features, such as support columns, walls, stairwells, and so on, with magic marker. Unnecessary features, such as wiring, plumbing, and so forth, are eliminated in this same step. Then the tracing paper is mounted on a board and photographed from different points of view. Once I have determined the best viewing angle, I enlarge the photo and trace off the perspective vanishing points, which gives me all the information I require to complete the drawing.

Examples of this technique can be seen on pages 130 and 142.

PAINTING WITH LIGHT

Some situations may call for photographic solutions that are a little trickier to execute. Assignments requiring pictures of objects with holes, slots, overhangs, or multiple layers, such as the automobile engine shown here, call for a more unusual approach, especially when very large or very small objects and poor lighting conditions are involved. For difficult subjects such as these, I employ a technique called *painting with light*.

Lighting a complex subject from a single source usually results in too many shadows; conversely, using a dozen or so floodlights to do the job would be too impractical. Painting with light avoids both problems by using a working exposure time of one to two minutes and a very weak light source—a 75- or 100-watt bulb mounted in an aluminum reflector does fine.

The camera is set on the tripod and loaded with 667 black-and-white film; the lens is set on bulb, the aperture to f/64.

All lights are switched off and curtains or shades are closed, making the general area as dim as possible. With the camera pointed at the dimly lit subject, the shutter is released and held open. I then switch on the low-wattage bulb and move it around in sweeping arcs and circles, pointing it into every nook and cranny of the subject, as if I were painting away all the shadows with a huge paintbrush.

I keep track of the exposure time with a stopwatch. When the exposure time is up, I turn off the light, walk over to the camera, and close the shutter. Because there is no way to determine exactly how long the exposure should be, I repeat the procedure, taking additional shots at different exposure times until I find the one that is exactly right.

Once I know the proper exposure time, I can take additional shots using a variety of painting moves, so I can identify problem areas that may require additional lighting time.

This image was achieved by painting with light, using Polaroid 667 film and an aperture setting of f/64 for a 45-second exposure. Generalized light painting throughout the image area left the back portion of the engine too dark, with loss of detail.

Here the 45 seconds of general light painting was supplemented with an additional 15 seconds of light concentrated on the back of the engine. Despite the maze of wires and belts, the result is a shadowless photograph with all details visible.

PHOTOGRAPHING THE EARTH

The technological advances of the past century have been astounding and unprecedented, outstripping the discoveries of all previous generations put together. We have even learned to look to the skies for the solutions to many of our earthly problems. As such efforts have accelerated, illustrations of the Earth as viewed from low-orbiting satellites have become increasingly important, proving valuable in studying such diverse subjects as population distribution, ecology, bird migration routes, missile deployment sites, geological features, and air traffic routes.

Since science illustrators tend not to find themselves in low-orbiting satellites too often, satellite photographs of the Earth are essential reference sources. In an assignment I undertook for *Scientific American* magazine, cloud formations obscured certain geographical areas and essential details on the satellite photographs I was referring to. Fortunately, I found I could get a much greater degree of accuracy by taking my own Earth photographs, without ever leaving my office.

My Earth was a metal globe, 12 inches (30 centimeters) in diameter and quite similar to the ones used in elementary schools. Using the same techniques described for the close-up model shots, I set about the task of photographing the globe.

The globe was prepared by painting out all the oceans with white acrylic paint and outlining all the land areas with a No. 2 black Rapidograph pen, thus making the land areas more identifiable in the photographs. A few latitude and longitude lines were left in place for use as reference points.

For photography involving such small-scale detail, I use Polaroid 665 positive/negative film. If an enlargement of only 300 or 400 percent is needed, I use the positive; for more substantial enlargements, I use the film negative.

This shot was taken with a No. 4 lens, and the camera positioned just 2 inches (5 centimeters) from the surface of the globe. Because of the depth of field the camera provides, it was possible to read the names of the cities near the horizon and pinpoint the smallest details with great accuracy.

Polaroid 665 film negative, shown greatly enlarged.

Polaroid 667 film positives, shown at actual size.

CHAPTER FIVE

THE PRELIMINARY STEPS— GATHERING INFORMATION

Clear communication between artist and client is the cornerstone of any successful creative endeavor, and this is particularly true of illustrating for science. The highly specialized and complex nature of the subject matter, the wide range of potential topics, and the need for absolute accuracy all combine to make it essential that the information relating to a given project be complete and thoroughly understood.

Illustrators can make things easier on themselves by asking the right questions of their clients, learning from previous assignments, discussing their projects with colleagues, and keeping up with current reference sources to stay abreast of developments in the various scientific fields. They should also keep in mind the various pressures and variables that can affect the other parties involved in a project—authors, editors, and art directors may have immediate priorities that differ from the illustrator's. Sensitive, reasonable concern for everyone's needs is the best way to achieve the common goal: a successful illustration.

Conceptual rendering for an article on atomic clocks.

THE ASSIGNMENT CONFERENCE

Once I've been commissioned to do an illustration, a conference is set up with the client and all other interested parties to discuss the design and general requirements of the proposed illustration.

While no two assignments are exactly alike, I always approach assignment conferences in pretty much the same way: My objective is to gather all the information I will need to do the illustration.

The assignment conference is best when limited to just me and the art director, but there are times when others, such as the account executive, editors, writers, and researchers, must be included. Generally speaking, the more important or complicated the assignment, the more staff I expect to deal with.

The meeting usually starts with a freewheeling discussion of the assignment's objective and what the art is intended to accomplish. This gives everyone an opportunity to voice their special interest in the project, and gives me a chance to see who is the technical expert and who will be responsible for approving the art.

Eventually the discussion focuses on the specifics of the illustration. I have discovered that the best way to get things moving is by making a quick sketch of what I think is wanted, asking questions and encouraging corrections and comments as I work. At this experimental stage, everything is in flux and is therefore kept simple—erase a line, make it longer, shorter, flatter, thinner, make a new sketch. When something is not completely understood, it is gone over until I am perfectly clear on every detail. Proposals are developed point by point, involving discussion on every aspect of what the final illustration should look like.

Although I may have a working knowledge of the particular subject, most assignments entail some technical language that is new to me. As unfamiliar terminology comes up, I

stop the conversation so it can be explained, which at the very least helps expand my own frame of reference on the subject, and sometimes even leads to an alternate approach to the illustration.

As I sketch and we talk, concepts begin to crystallize, and what initially may have seemed to be a complicated job now begins to simplify. We may see if it works as a single schematic drawing, or perhaps by separating the elements into easily followed modules. Or we may end up adding additional views to enhance the main drawing. Color is discussed and noted on an overlay.

I make notes on everything, including where to obtain additional reference material, should it be needed. If photographs or models are supplied, they are systematically numbered and placed in protective folders. Toward the conclusion of the meeting, other factors are discussed and resolved, such as deadlines, type, art style, and budget. I also take notes concerning sizes, proportions, materials, and any other technical information that might apply to the given situation.

By the end of the meeting, most of the problem areas have been dealt with and I have a rough pencil sketch showing all the elements necessary to the illustration.

Of course, not all assignment conferences go as smoothly as this. For example, if it appears during a meeting that contradictory or inconsistent answers to my questions are coming from several different people, it is often best to recess the meeting until definitive information can be gathered and clarified. Sometimes the project is of such a technical nature that the staff has not anticipated (and is therefore not equipped to answer) all my questions, in which case the author or a technical expert must be consulted. If the expert is not close by, the conference sketches may have to be faxed back and forth several times until all the details can be worked out.

Bank of computers transmitting information into a fiber-optic cable network. The transfer of information between authors, editors, and illustrators may not be as fast, but it must be just as precise.

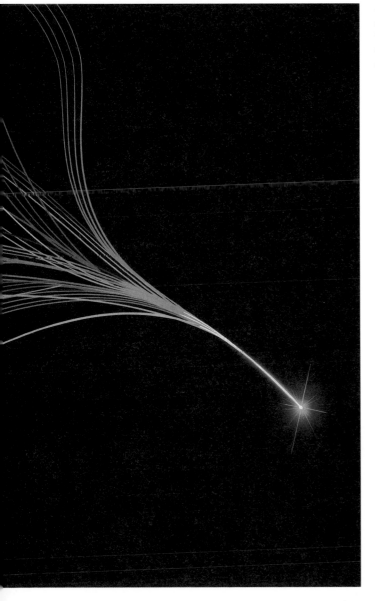

WORKING WITH AUTHORS

The author's responsibility is to communicate the requirements for the illustrations that are to accompany the text; the artist's responsibility is to help the author accomplish this task with as little confusion and misunderstanding as possible.

Most author–artist communiqués are routed through the art director, thereby allowing the art director to stay in touch with the development of the illustration and maintain an overview of the project. Particularly difficult problems, however, are handled through direct author–artist contact, either by phone or in person.

In my experience, only a handful of authors have been able to draw well enough to supply me with adequate hand-drawn references. More often they clip drawings from other books, frequently giving me three or four references for a single drawing. Sometimes I am asked to extract selected portions from each reference source and combine them to design a new drawing. Since each of the drawings originates from a different source and artist, the scales and points of view are unrelated, making it very difficult to work in this manner.

All the reference drawings and photographs used for the lunar observation base illustration on page 81, for example, were originally done for other purposes, from differing perspectives, and at varying sizes. Combining them into a composite image required two preliminary drawings. First a pencil layout over a perspective grid established a basic scale of one grid square equalling 4 feet (1.2 meters). All reference components were then redrawn and sized to fit into the landscape environment. Once these basic spatial relationships were finalized, a second, more finished drawing was prepared.

One good approach to author–artist communications is through an author-supplied *operational box figure*, similar to a flowchart, showing the relationships of the illustration's components, plus a written description identifying the purpose and function of each box. A rudimentary drawing of this type, along with any general reference material the author can furnish, will usually provide enough information to make a preliminary drawing.

Occasionally an author asks to have every bit of detailed information discussed in the text incorporated into the illustration. It serves no purpose to argue about these things—I can well understand how anxious the author might be to see his or her work published exactly as it was envisioned, complete with unedited text and data-packed illustrations. But sometimes, as Mies van der Rohe said, "less is more," and this can be easily demonstrated to even the most stubborn author. Working at two to three times larger than the illustration's final size, I do a rough "convincer" sketch, incorporating every author-requested detail and nuance plus the corresponding labels. Then I photocopy it at the proper reduction and produce a final-size copy, which is sent off to the author along with the original. This tends to end the discussion—it is immediately apparent to the author that the publisher would have to distribute magnifying glasses in order for the illustration to be of any value.

All of which just goes to show that a picture really is worth a thousand words.

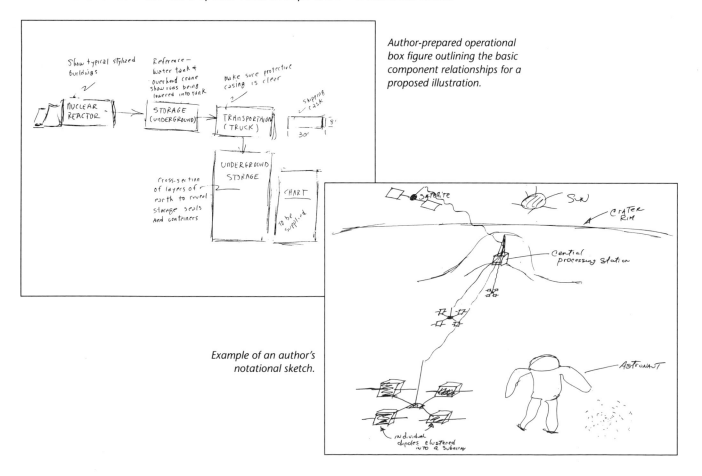

Author-prepared operational box figure outlining the basic component relationships for a proposed illustration.

Example of an author's notational sketch.

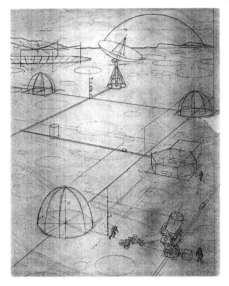

First preliminary drawing for lunar landscape illustration. The two vertical scales used to size structures can be seen near the moon vehicle in the foreground and near the radio telescope in the background. Random ellipses show placement of craters. The Earth was drawn at an arbitrary size on the horizon.

Final illustration, incorporating the author's modifications to equipment and landscape. The author also supplied the calculations needed to determine the proper size of the Earth.

Second preliminary drawing. In this more finished version, the scene is illuminated as if the sun were near the lunar horizon, with shadows plotted to fall at correct angles and lengths. This drawing was shown to the author.

WORKING WITH EDITORS

Although the author is the expert and final authority for any given project, it is the editor who controls the entire project and is primarily responsible for its smooth progression. In addition, when the author is not available to discuss an assignment, it is the editor who steps in, acts as the intermediary, deals with problems that arise, and ultimately sends the pencil sketches to the author for approval.

The best editors think ahead about the artist's needs, providing all the necessary research material and keeping lines of communication open between themselves, the author, the art director, and the artist. The sharpest editors are also the pickiest, running themselves ragged to check out the smallest detail on a reference.

Because we live in a fast-paced world where dramatic technical and scientific advances occur almost daily, science editors face a monumental task staying abreast of the new developments in so many areas. In planning a project, editors customarily gather the best reference material available on the subject and later use that material as the primary guide for cross-checking the accuracy of the art. Sometimes, however, rapid scientific developments effectively make references obsolete almost as soon as they are printed. In other cases, the reference material may contain subtly arbitrary or misleading information too technical or esoteric for the editor to notice. Naturally, these problems get passed on down the line and become problems for the illustrator as well. It is therefore essential that illustrators be alert for such errors.

In one such instance, I was to illustrate a precise sequence of base pairs on a DNA chain (see preliminary sketch below). Although my instruction from the editor was to follow the sequence exactly, my experience with a considerable number of DNA illustrations led me to conclude that the coding sequence was probably random. The editor remained unconvinced, directing my attention to the piece of reference art he was using as a guide. The art turned out to be a piece of my own work and actually reinforced *my* position. Scientific accuracy aside, it is important to see how this example demonstrates the artist's responsibility to ask questions and exercise professional judgment, rather than just blindly following directions. Anything less effectively reduces the illustrator into being a technician instead of an artist.

A face-to-face meeting with the editor is the best way to guarantee a clear and accurate transmission of information. I like to sketch on the spot and get direct answers to my questions. For those who think I may be overemphasizing communication skills at the expense of drawing skills, consider the demonstration on page 83, which illustrates the variety of interpretations that can be made from one drawing.

Preliminary pencil sketch of a DNA double helix.

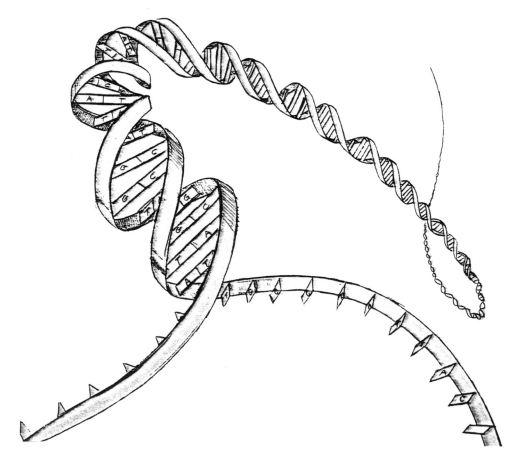

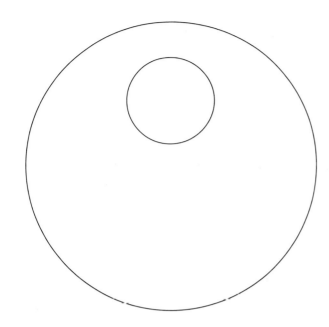

QUESTION: Is this

A) a cardboard disk with a hole in it?

B) two circles of wire?

C) a view through a long pipe?

D) a view down a cone?

E) two spheres floating in space?

F) a cam on an axle?

ANSWER: It is nothing, _until you ask the right questions._

PROFESSIONAL TIPS

For the uninitiated, an art conference with an author, editor, or other project staff members can be a tense experience. These few suggestions may serve as a guide to ease the situation:

1. Dress properly—jacket and tie. While less formal attire is acceptable for staff artists, it is not appropriate for a businessman attending a business meeting, which is what an art conference is.

2. Come prepared to work. Bring a small, well-organized briefcase with your sketch pad, pencils, note pad, overlays, and whatever else you think you might need. Do not expect the client's art department to supply you with materials.

3. Do some homework beforehand. When possible, do some preliminary research on the subject matter and bring the reference material with you for discussion.

4. Formulate a list of "need-to-know" questions. "When is the deadline for the final art?" is a classic example of a question that seems so elementary that you may forget to ask it unless you have it on a prepared list.

5. Take careful and complete notes; trust nothing to memory.

6. Do not waste time; keep casual chitchat to a minimum. Everyone else at the meeting has other things to do, and so do you.

7. Determine who is the technical expert on the project, and who has ultimate responsibility for the art.

8. Above all, ask questions—lots of them. Be sure that each one has a satisfactory and complete answer, and don't leave the meeting until every one is resolved.

MAKING A PRELIMINARY PENCIL SKETCH

In a sense, an illustration starts when the assignment conference ends. I spend the trip back to my office mentally reviewing the assignment and planning my approach. By the time I reach my studio, I usually have a fairly good idea of what the final image will look like.

The preliminary pencil sketch is the key stage, the moment of truth, for an art assignment. Before starting a drawing, I devote a significant amount of time to reviewing the notes, rough sketches, and data I have compiled from research and conferences, checking the accuracy of the reference material down to the smallest detail. If a model is needed, this is when I construct and photograph it. Only when I am satisfied that everything has been thought out and accounted for do I sit down to draw.

For the preliminary sketch, I use a mechanical pencil with a 3H lead on tracing vellum, which produces a smudge-free, reproducible line drawing. To suggest color, I either photocopy the drawing and then add various tints or indicate my color choices on a tracing paper overlay, both of which allow color changes to be made later on without disturbing the original drawing.

I believe in planning and executing preliminary sketches with the same attention I give to final art, so I am a stickler about drawing in every detail of every shape, even at this early stage. I never take shortcuts or try to save time by leaving out small details. Experience has taught me the prudence of proceeding slowly and the value of having the work checked at every stage.

Submitting carefully executed drawings for approval, rather than rough, less finished sketches, precludes the possibility of the client assuming that the final art will include anything not clearly apparent in the preliminary drawing. I also make every effort to have the project expert initial the sketches—asking for an approval signature in this manner ensures that the reviewer will give the art his or her undivided attention. Once the pencil sketches are approved, I can complete the final art with reasonable certainty that it will meet all the criteria. In the final analysis, the extra time spent on painstakingly prepared preliminary art is justified by the time saved on extensive corrections and revisions—when I submit final art, there are no surprises, no questions, and rarely any corrections.

Finally, there is also a psychological factor to consider: A job that has been handled carefully and professionally from start to finish reflects positively on the art director's decision to have hired you for the assignment in the first place. An art director who has been spared the typical hassles and headaches on a project is an art director who will probably be calling you with more work in the near future.

Working-size preliminary sketch of a nuclear reactor in its containment building, with the heat-exchange building to the right.

CHAPTER SIX

CREATIVE PROBLEM SOLVING— SHORTCUTS, SECRETS, AND TRICKS

In the course of producing illustrations, even the most experienced artist will be confronted by problems that cannot be solved by conventional methods. This may be due to the physical parameters of the problem, the complex nature of the scientific phenomena involved, or a variety of other reasons.

The key to solving such problems lies in staying focused, maintaining one's resourcefulness, and having faith in one's own creative ability to solve any problem at hand. Given enough time, thought, and imaginative energy, virtually every problem yields to skilled illustration techniques. The following are some suggestions for getting through those difficult times, along with some of my own secrets and tricks for producing interesting special effects.

Cover illustration based on a mathematical theory analyzing the shape of knotted curves in three-dimensional space. Although the twists and tangles of this lengthy loop of rope create an intricate maze, they do not form any knots.

PAINTING A PERFECT LINE: THE BRUSH STYLUS

Painting a fine, highlighted edge on an ellipse or curve is risky to attempt freehand. When a series of repetitive curves must be highlighted, the odds of duplicating each edge exactly are remote. Even straight-line edge definition can be a challenge when the lines are 10 inches (25 centimeters) long or more.

Working with a fine brush and using a ruler or French curve as a bridge guide is one of the more typical approaches. But this requires a great deal of practice, and even then it is not very reliable, especially if the line must be retraced in order to build more density.

Ruling pens tend to bleed and run paint, often leaving blobs at a line's beginning and end. In addition, they are poorly suited for rendering thick lines and impossible to use for tapered lines. Except in the hands of the very skilled, they tend to cut or scratch as they are drawn across the surface of the illustration board. Wrico pens are a bit better, but the capillary point limits their usefulness for finely drawn lines.

Steel-tipped instruments such as ruling pens and wrico pens ride on the peaks of the texture of the board, leaving ragged edges, whereas the soft, flexible tip of a paintbrush can be adapted to ride easily over the irregular surface, which brings me to the most important art instrument in my studio: *the brush stylus.*

Using this relatively simple tool of my own design, I can do precise brush line work that would be impossible to achieve by any other method. The illustrations on this spread show how to use the stylus and the variety of effects that are possible with it.

Since a large part of airbrush work involves highlighting and edge definition and there is no commercially available instrument well suited for this type of delicate work, the best approach is to construct your own brush stylus. Start by taking the dowel from a wooden coat hanger and cut it to a 7-inch (17.8-centimeter) length. Use a mat knife to carve a V groove at one end, about 4 inches (10 centimeters) long and halfway into the dowel's cross section. Continue to shape the groove until it accommodates the shaft of a series 7 sable brush No. 2, and then sand until smooth. Gently tap a 1-inch (2.5-centimeter) finishing nail into the grooved end, and finish the wood with a coat of clear nail enamel, which will help keep it clean and smooth.

To use the stylus, place the brush in the V groove so that only the very tip of the brush contacts the illustration board. Placing the finishing nail against a straightedge or French curve, draw the stylus along the edge of the guide. The stylus will enable you to lay down a perfectly straight, curved, or tapered line as thin as a hair or as much as a quarter of an inch (64 millimeters) thick.

An additional tip for anyone thinking of trying this technique: Work with a runner of acetate under the edge of the ruler or French curve guide. This will prevent the nail head from coming into contact with the surface of the illustration.

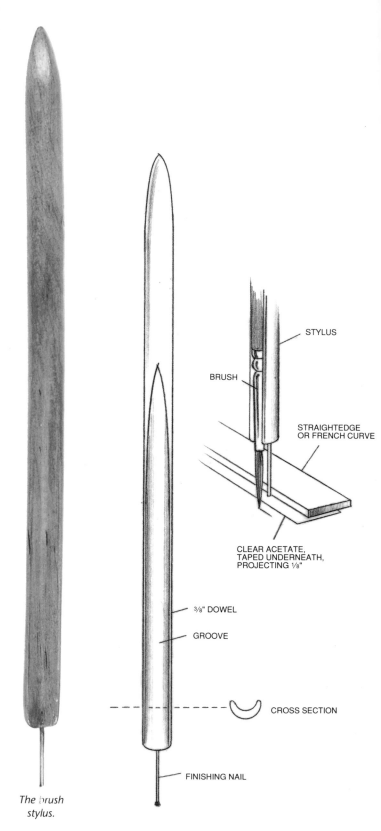

The brush stylus.

STYLUS

BRUSH

STRAIGHTEDGE OR FRENCH CURVE

CLEAR ACETATE, TAPED UNDERNEATH, PROJECTING 1/8"

3/8" DOWEL

GROOVE

CROSS SECTION

FINISHING NAIL

Construction details for the brush stylus.

Using the brush stylus.

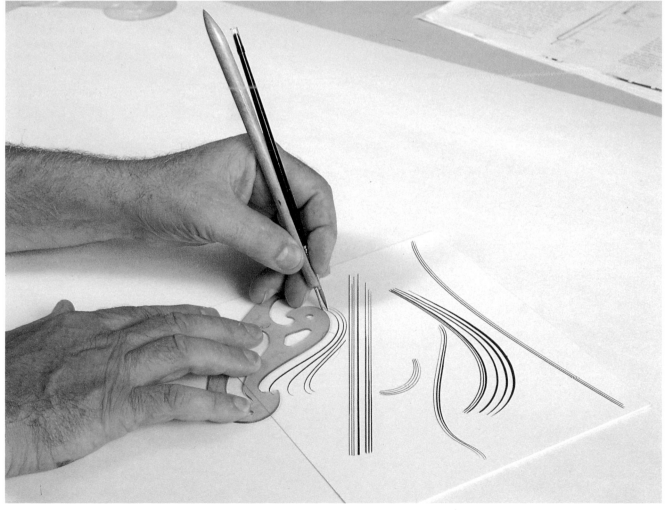

DRAWING CURVED PARALLEL LINES: THE SHOULDER SPACER

If you have ever tried to draw a set of perfectly parallel lines in an elliptical, curved, wavy, or some other irregular shape, you have discovered that it takes a little ingenuity to accomplish. Techniques that are adequate for drawing sets of short parallel lines are not necessarily adaptable to more intricate work.

Simple parallel lines can generally be "eyeballed," or a line can be drawn and perpendicular points measured out from it to establish the second line. Another commonly used method calls for inking a "fat" line and then tracing each side of that line. Those with an adventurous nature may have tried using a draftsman's double-pointed railroad pen with a swivel head.

These methods' inherent drawbacks, however, make the techniques poor choices for more complicated shapes. The double-pointed pen, for example, can be used with a French curve guide to draw gently curved lines but is ill-suited for drawing tight, wavy lines.

Since the need for parallel lines occurs regularly in technical art, I found it necessary to design something that would produce a consistently perfect set of parallel lines regardless of their shape or complexity. By approaching the problem from a new perspective, I came up with what has proven to be the perfect solution: *the shoulder spacer*.

To make your own shoulder spacer, start with two Rapidograph pens of equal size and some masking tape. Cut a 2-inch (5-centimeter) length of masking tape, approximately one-sixteenth of an inch (16 millimeters) wide. Begin by wrapping the tape around the point of one pen, just exposing the tip of the point. When the tape has built up to the point where its thickness equals the distance to be maintained

between the two parallel lines, cut off the balance of the tape. When using the pen, the masking tape forms a "shoulder" that keeps a consistently uniform distance between the point of the pen and the French curve or ellipse guide.

To use the shoulder spacer with an ellipse guide, first raise the guide off the surface of the board by placing a slightly larger guide underneath it. Ink the first of the parallel lines with the *un*wrapped pen. Without moving anything, ink the second line using the pen *with* the masking tape shoulder. You will now have two perfectly parallel elliptical lines.

The same technique works equally well with French curves, which need not be lifted off the art surface since they are several times thicker than ellipse guides.

For more complicated parallel curves, such as those in the illustrations shown on page 91, the basic method remains the same, but it is best to construct your own guides, as follows: First, cut all the curved lines as accurately as possible into a sheet of .005 acetate; after cracking the acetate, smooth off any rough spots with fine sandpaper. With the acetate functioning as a guide, use a pencil to trace its shape onto thin, soft cardboard, such as shirt cardboard. Then use an X-acto knife to cut out the cardboard shape, cutting one-sixteenth of an inch (16 millimeters) in from the pencil line at all points.

Spray the cardboard with adhesive and place the acetate on top of it. The acetate will now be high enough off the surface of the art to allow the taped shoulder on the pen to ride against the acetate edge. Ink the lines as explained previously—the first line with the untaped pen, the second with the pen that has the shoulder spacer.

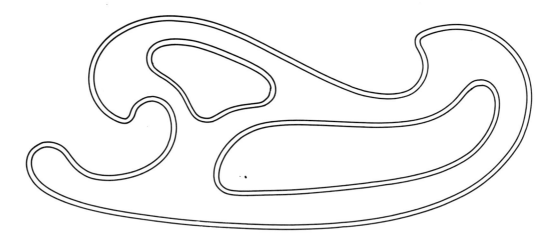

Parallel lines drawn around a French curve.

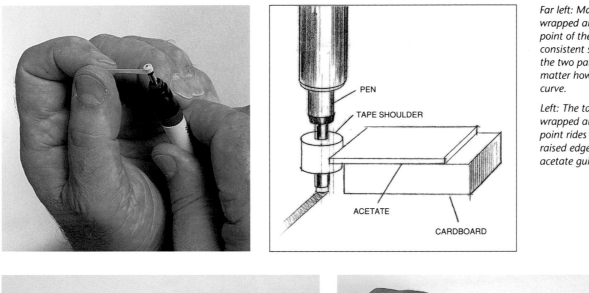

PEN

TAPE SHOULDER

ACETATE

CARDBOARD

Far left: Masking tape wrapped around the point of the pen creates a consistent space between the two parallel lines, no matter how much they curve.

Left: The taped shoulder wrapped around the pen point rides against the raised edge of the acetate guide.

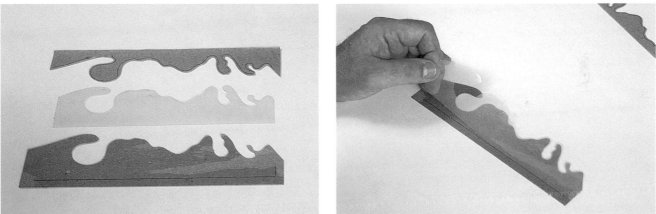

The acetate guide is mounted on thin cardboard to raise the acetate edge high enough for the taped shoulder to ride against it.

Variety of parallel lines executed with a shoulder spacer.

MASKING FOR AIRBRUSH RENDERING: ACETATE .005 CLEAR

Prepared frisket paper has been the most commonly used masking material for the last 30 years. Before masks can be cut, however, the drawing must be transferred from a tracing paper layout to illustration board. The frisket paper is then laid down on the board and the masks are cut out with an X-acto knife.

There are a number of disadvantages to this approach. To begin with, it is almost impossible to avoid cutting through the frisket paper to the surface of the illustration board itself. Frisket paper is also very difficult to lay down on large areas and, once positioned, is not easily lifted and repositioned. In addition, the paper's tendency to wrinkle and stick to itself makes it impractical to save for future revisions. There is a much better material to use: *acetate .005 clear*, which is easily cut or scored and can be lifted and repositioned as often as necessary to check the progress of the work.

A

In the course of laying out a job, all my masks are cut out of this acetate and pinhole register marks are put in each of the corners, so that each mask to be positioned on the board is in perfect register with the pin marks. There is no need to transfer the pencil drawing to the illustration board, since the masks themselves *are* the drawing. More complicated illustrations require more masks, but acetate, unlike frisket paper, can be easily raised above the board surface, making special airbrush effects possible. And there is no problem in saving the masks should corrections become necessary.

To cut straight or curved lines in acetate, use an X-acto knife and lightly cut the surface. To score ellipses or circles, use an engraver's needle or the needle point on a compass and, placing an ellipse guide on the acetate, gently scratch a line around the guide. For ellipses or large curves, make a single or double set of scratch marks across the curve to act as alignment guides, which help in repositioning the pieces during the course of the work.

B

Once all the shapes have been cut or scored, turn the acetate over and lightly spray the back of each piece with Scotch Spray Mount Artist's Adhesive No. 6065. The individual pieces can be separated by gently flexing the acetate back and forth. The adhesive's light tack will allow you to lay out all the pieces on paper in the order you expect to use them and still permit you to lift them up easily and reposition them as needed. At the end of the job, the masks can be placed on paper and filed away in their original positions in case corrections are needed at a later date.

Only two basic acetate masks were used to create this comprehensive cover proposal on the science of robotics. Once the masks were cut and separated, the individual parts were used to define the figures and to mask out the sections I did not want painted. Clear acetate does not photograph well, so in order to reproduce it for this illustration, I sprayed it a neutral gray. Note the pin registration marks, which appear as black dots in the four corners of the master acetates (A and B).

The following sequence was employed:

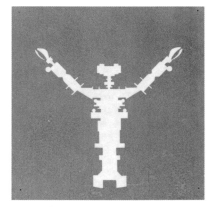

C

1. Mask A was positioned on a four-ply Strathmore bristol board. A blue-black background was sprayed on, and then a lighter background tone followed for the head and chest areas.

2. Mask A was removed and replaced in the same position with mask B, which was sprayed an even lighter blue. The area intended for the robot's head was sprayed the lightest shade. Pin marks, piercing through to the board, were placed in the four corners of the mask.

3. Now B was removed and replaced with C, which was positioned in perfect pinhole register. The robot was sprayed with flat ivory black, and blue-gray was used to pick up some suggestion of detail.

4. Some simple line work and the addition of the red circle and eyes completed the illustration.

Note that D, which was the center section of C, was not used. It is included here to show that I had the option of using several other versions of this concept—acetate masks allow for many variations of a theme. They also permit you to monitor the progress of the work, since they can be easily lifted and repositioned.

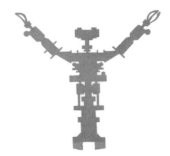

D

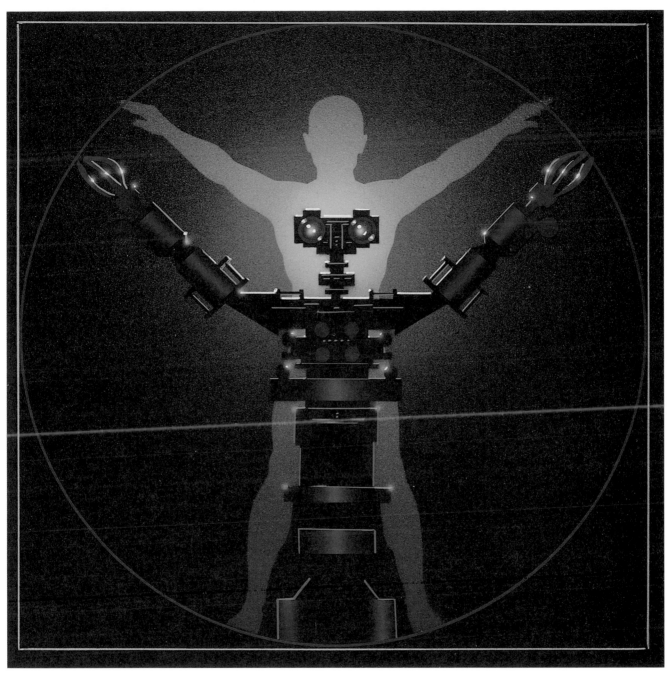

Final comprehensive color illustration.

GOING FROM HARD EDGES TO SOFT: LIFTING THE MASK

The edges on airbrush renderings of mechanical objects tend to be hard and knifelike, while for biological subjects they are usually soft and rounded. This general rule of thumb, however, may not always apply—there are times when you have to show a smooth transition from one type of edge to the other. A variety of transitional effects are possible by *lifting the acetate mask* above the surface of the illustration.

A view of a cell's nucleus encased in a fluid environment,

for instance, would be expected to have nice, sharp edges in front and somewhat diffused, fuzzy edges at the rear. To mask properly for these effects, first apply spray adhesive to the back of the mask. Stick one end of the mask to the surface of the illustration while you lift the other end slightly. The higher the angle of the mask, the softer the transition will be. Airbrush until the desired effect is achieved. For greater variety, the mask can be bent or curled.

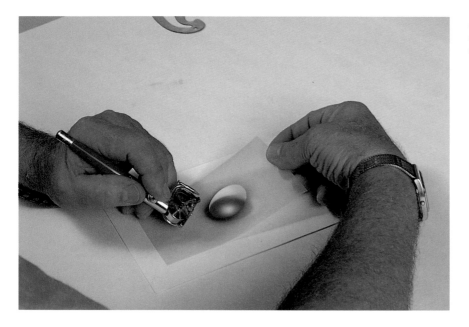

Lifting one edge of an acetate mask and airbrushing through a circular cutout, and the resulting soft-edged shape.

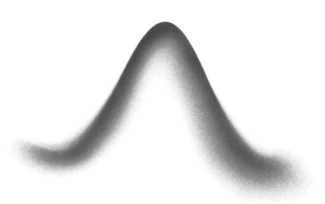

Curling an edge of an irregularly shaped mask, and the resulting hard-to-soft-edged effect.

Rendering Repetitive Shapes: Indexing

In many scientific disciplines, particularly molecular biology, the problem of painting repetitive shapes frequently confronts the artist. A time-saving method for dealing with this issue should be a worthwhile addition to the repertoire of any scientific illustrator.

To prepare masks for an airbrush illustration that repeats a pattern of molecules running 30 wide and 10 deep on a plane surface, as shown here, would take many hours of preparation. But by cutting two relatively simple masks designed for *indexing*, a difficult procedure can be made easy.

To create the illustration on this page, I began by cutting 15 circles out of a sheet of .005 clear acetate. I airbrushed the last 10 circles, leaving the first five untouched. Lifting the acetate mask, I repositioned—or indexed—it to the right,

placing the first five unused circles over the last five airbrushed ones. This correctly positioned the next 10 circles for painting. I continued indexing in this manner until all of the molecules in the first row were rendered.

As you can see, the second row of molecules is partially covered by the first. A second mask was cut to represent the change in shape caused by one row being partially obscured by the other, and the same indexing procedure was followed.

Without using indexing, this illustration would have taken three or four hours; with indexing, however, it took about an hour. Once the technique is understood, you will discover other variations and applications for it. A more complex example can be seen in the illustration of the AIDS virus on page 119.

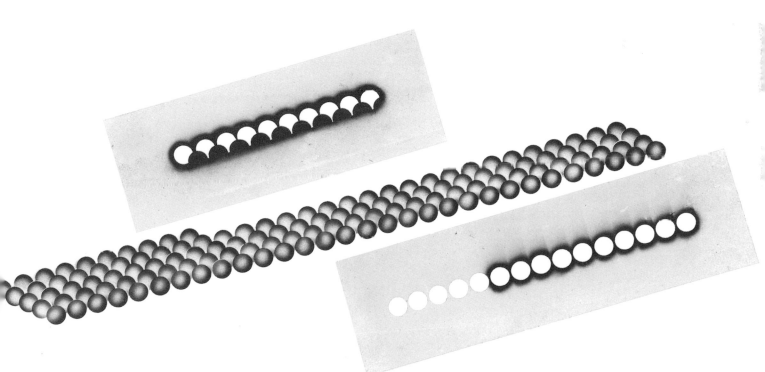

Using two masks—one for the front-row molecules and another for the partially obscured molecules in the second and third rows—allowed the quick creation of dozens of identical images.

CREATING SOFT-EDGED, CONCENTRIC CIRCLES: THE TURNTABLE

Electrons orbiting around an atomic nucleus form concentric shells of varying width. For an illustration showing the relationship of the different shells, a cross-sectional view was selected. For the orbits to appear soft and glowing, the circles would have to be drawn with the airbrush.

In considering the suitability of my airbrush masking technique, I realized that in this instance masking presented a unique set of technical problems that would not be easily solved. To get the soft-edged effect I wanted, the masks would have to be raised above the surface of the illustration board with the masks positioned one inside the other and a consistent space maintained between them. But I could not find a way to support the masks and keep them in register at the same time—another way had to be found.

Fortunately, I am a collector, almost to a fault—I like to save the kind of things that some might call junk. In my basement workshop, home to a wonderful assemblage of mechanical artifacts and assorted pieces and parts, I often find just what I need for creative solutions to difficult problems.

Rummaging through my collection, I came across a *turntable* from an old record player. I mounted the turntable's platter on a wooden base and, centering my illustration board onto the platter, taped the one onto the other. By holding the airbrush at a fixed point and spinning the platter with my other hand, I was able to blow perfect soft-edged, concentric circles.

Airbrushing circles onto illustration board mounted on spinning turntable platter.

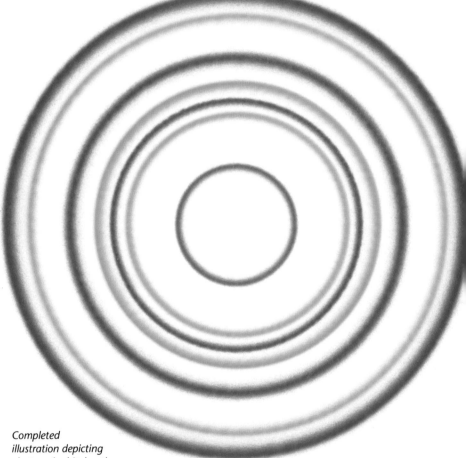

Completed illustration depicting electrons' orbital paths.

ACHIEVING TEXTURED EFFECTS: SIDE SPRAYING

Most artists are familiar with the variety of creative textural effects that can be achieved by dry brushing, sponge, cloth, and spattering techniques. *Side spraying*, the method presented here, offers some interesting new possibilities.

Prepare the surface to be textured by brushing on a heavy layer of white acrylic paint. Create a stuccolike texture using a flat piece of acetate, daubing it at the paint and lifting it quickly. With this stucco texture serving as a foundation, use a toothpick to do additional modeling and to suggest lines and hatching patterns. Try a stiff brush instead of a pick to develop still other effects. After a little practice, experiment with screen, cloth, fiberglass, gauze thread, and anything else that might yield an interesting texture.

A little shrinkage can be expected to occur as the acrylic surface dries, but there will still be enough raised texture to airbrush a whole range of exaggerated effects. Use designer's gouache to spray the entire textured surface with a basic background color and, working from an extremely low angle, almost parallel to the surface, blow in a shadow color, darker than the background, onto one side of the acrylic texture.

You will find that even a slightly raised textured surface appears to have great depth when sprayed on one side from a low angle. This method was used to produce the enhanced textures shown on this page. The technique has many applications in the Earth sciences and is a nice way to enliven the background areas of charts and graphs.

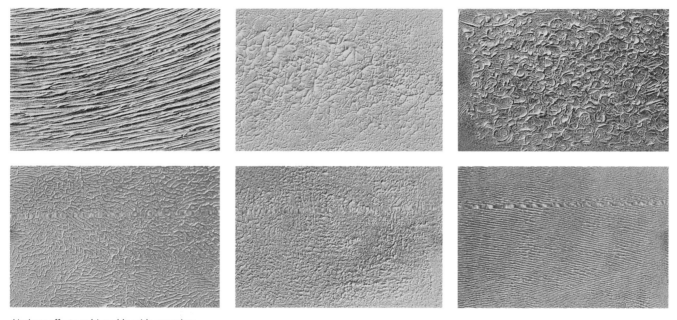

Various effects achieved by side spraying.

ADDITIONAL AIRBRUSH TIPS

Some additional tips regarding use of the airbrush will aid in solving some common but tricky problems. For instance, even the most careful artist faces the disaster of having the airbrush spit or spatter from time to time. I try to minimize the risk by keeping all areas I am not working on covered with either acetate or tracing paper and working on only one small exposed area at a time. Should the airbrush spit or spatter, immediately blot up the excess paint with a tissue and dry the area as quickly as possible. Trying to hide the spatter by applying more color over it is a waste of time—the separator's color scanner will see through it and pick up the spotting. Instead, to salvage the painting, spray the damaged area with opaque white before continuing to paint.

For painting large backgrounds, I use a single-action airbrush, which puts out a lot of pigment and spray very rapidly. This kind of heavy spraying causes some of the pigment to dry before it actually hits the surface of the painting, resulting in the accumulation of a fine layer of pigment dust on the

surface of the art. This pigment dust can be prevented from contaminating surrounding colors and causing streaking by spraying the painted surface with a solution of one drop of gum arabic dissolved in 10 drops of water. This mixture stabilizes the paint by binding the pigment to the surface. Use a No. 5 tip on a single-action airbrush to regulate the spray.

When the work has been completed and all the acetate masks have been removed, the surface can be safely wiped with a soft tissue and thinner to remove any residue from the spray adhesive that may have been left by the masks.

Since geology books in particular usually call for repeated use of the same color combinations throughout the course of a project, mix large amounts of the necessary colors and store them for later use. I store mixed paint in the plastic containers from 35mm film—they're airtight and keep the paint from drying out.

Finally, when large background areas need to be dried quickly, a common hair dryer does the trick quite nicely.

SHADOW-IMAGE PAINTING

These familiar chemistry test tubes take on a whole new appearance after rendering with the unusual masking technique for *shadow-image painting*. In this instance, the glass apparatus was laid flat on illustration board. A background airbrush was used to spray the various colors from different angles with the glass tubes acting as masks. When the tubes were removed, their soft shadow images remained.

It is possible to achieve a wide variety of interesting effects this way: Colors, alone or in combination, can be sprayed over the surface of the illustration board prior to positioning the glass tubes; contrasting colors or color combinations can be used once the tubes have been laid down; varying the height of the airbrush spray also produces some surprising effects as the paint flows around the curved surfaces of the tubes.

I have gotten some equally beautiful results by spraying over modeling clay, rocks, driftwood, and a number of similar items. The ethereal effect will be different with each object. Shadow-image painting works particularly well when used in combination with typography, and produces wonderful backgrounds for annual reports and brochure covers.

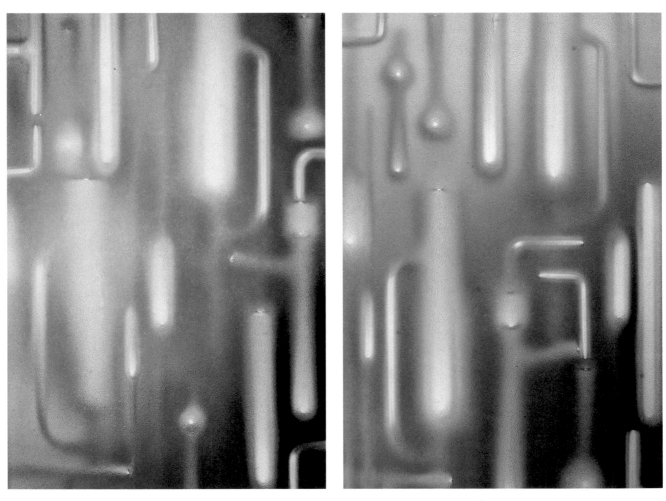

For shadow-image painting, the actual objects to be depicted—chemistry test tubes in this case—can be used as masks, resulting in soft, handsome effects.

COMBINING ART AND PHOTOGRAPHY

A magazine assignment found me looking for a way to show the significant role chemistry plays in the production of modern, everyday products. My original idea was simply to show an array of product photos against a background of glass test tubes and beakers. I had started on the first pencil sketch when it occurred to me that it might be more interesting if I could place the photographs of the products *inside* the glass tubes.

I called the magazine's production department to see if what I had in mind was feasible. Together we were able to work out the technical details. My rendering of the glass tubes was confined to the outside edges and kept to a minimum. The inside of the glass had to be almost pure white so that the photos could be masked and overprinted on the art. I used a stucco-textured background to provide some contrast to the glass tubes. The final printing produced an attractive blend of photography and art.

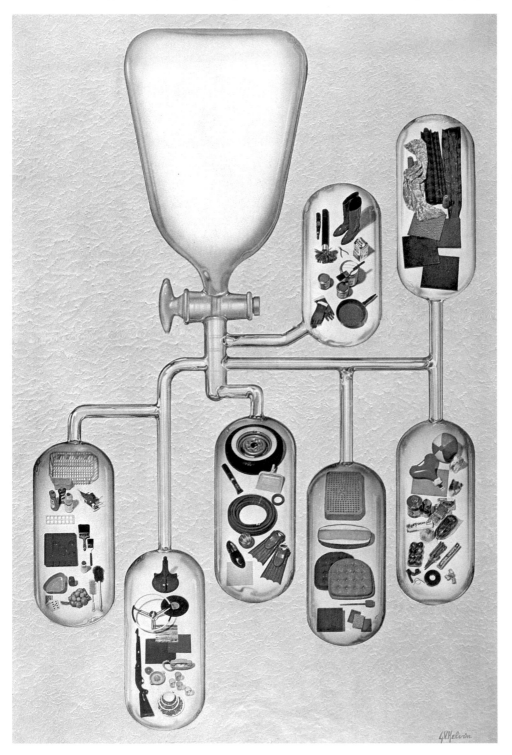

A combination of photographic and artistic techniques coupled with severe alteration of scale allow this illustration to make its point on the value of chemistry in everyday life.

PROVIDING A FOCUS

Some subjects are so complicated that they threaten to overwhelm the reader completely. Fortunately, there are design techniques the artist can use that keep the integrity of the different components intact while making things easier for the reader to follow.

While every element is important to a whole concept, complex subjects usually feature at least one section that deserves special attention. Once I determine (sometimes in consultation with the editor and/or author) which element should be isolated and highlighted in this manner, focusing on the special area helps to establish the relationship of the different elements.

There are a number of ways to establish this focus, some less imaginative than others but effective nonetheless. These include using indicator arrows, circling or boxing specific areas, using tints to distinguish one element from another, or any combination of these. When audience sophistication and my color options permit it, I employ more subtle techniques, such as color against line or color against a neutral background; if I am limited to line only, a black line against a slightly grayed or colored line works well, as does a thick line against a thin line.

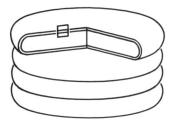

This portrait of a rhodopsin molecule embedded in a cell membrane focuses attention directly where it is needed—on the molecule itself. The cell membrane is shown in black line and only for the purpose of establishing a contextual environment. The blue DNA helices are kept pale, so the red central molecular structure can easily be seen even though it is small when compared to the entire molecule. The diagram at top left shows the location and scale of the section.

CYTOSOLIC SIDE

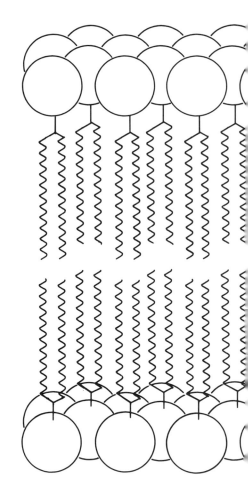

INTRADISCAL SIDE

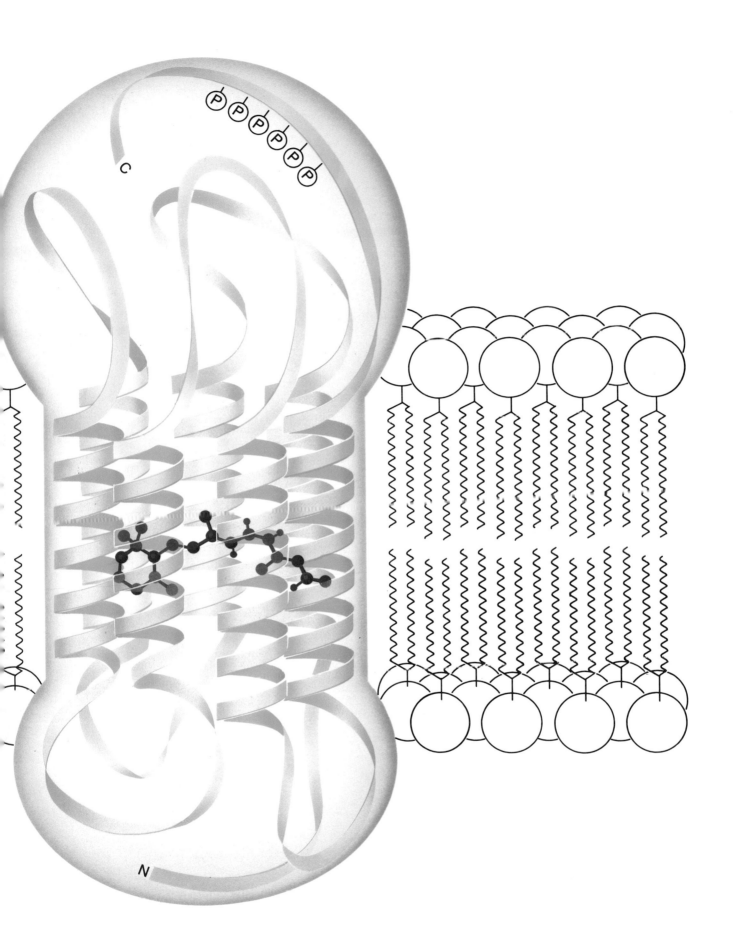

TRIMETRIC PROJECTION

Projections are methods of drawing that employ mathematical formulae. Their primary advantage is that they allow direct measurements from an object to be easily converted to a drawing with a simplified reduction scale, which can be particularly useful when rendering anything requiring a consistent viewing angle, such as an exploded view.

There are many projection techniques available to the skilled artist, but I prefer *trimetric projection*, which creates the most realistic image. In the trimetric process, all lines in each plane are parallel and there is no convergence as there would be in a perspective drawing. Instead, there are separate foreshortening scales for the right, left, and vertical axes, which combine to give the drawing a natural appearance even though the lines do not converge to a vanishing point.

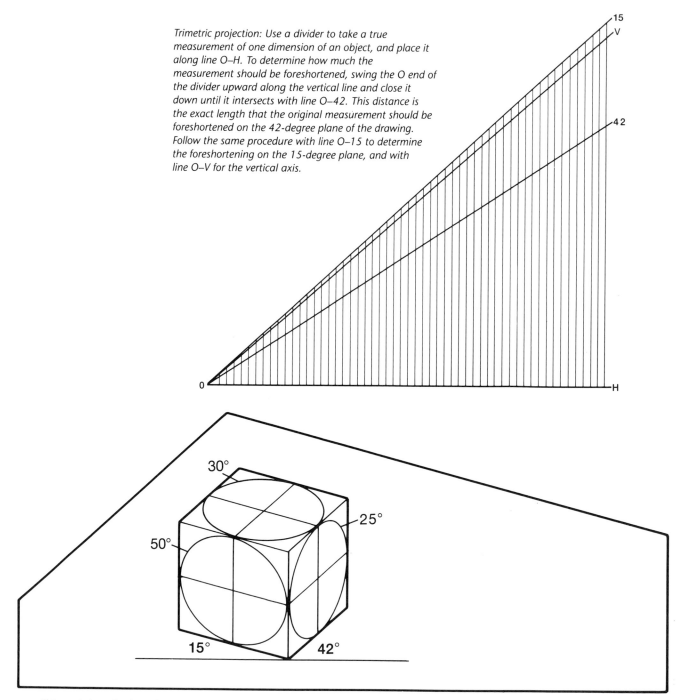

Trimetric projection: Use a divider to take a true measurement of one dimension of an object, and place it along line O–H. To determine how much the measurement should be foreshortened, swing the O end of the divider upward along the vertical line and close it down until it intersects with line O–42. This distance is the exact length that the original measurement should be foreshortened on the 42-degree plane of the drawing. Follow the same procedure with line O–15 to determine the foreshortening on the 15-degree plane, and with line O–V for the vertical axis.

This rendering of a three-dimensional trimetric guide provides the exact angles needed for producing trimetric drawings, showing the proper ellipses for each plane. Note that the lines for each face of the cube are parallel. The cube drawn on the actual guide is tipped toward the viewer at a 30-degree angle, so that the top face is visible; the 15- and 42-degree angles remain constant.

Color and Legibility

Judicious use of color can help in communicating the illustration's point to the viewer. Color has three dimensions—value, hue, and tone—and these, in combination with shapes, character, texture, and purpose, influence artistic color choices. There are a number of ways that color can be manipulated to produce certain results: Dramatic effects can be achieved by working dark against light and vice versa; moods can be conveyed with somber colors; light, airy effects can be created with high-key pastels. Establishing the background color first enables me to balance the color relationships between the primary subject and its environment; working in reverse—rendering the subject first—tends to make the components look as if they were pasted onto the background.

When designing an illustration in segments, assigning a uniquely identifiable color to each component helps the readers to track them as they reappear in various sections of the piece. These colors should be as close as possible to the primaries, to make them easily distinguishable from surrounding shades. Tints of similar values should not be used in the same illustration, especially not in close proximity to each other.

While there are no "right" or "wrong" colors for most situations, color standards already exist for some scientific disciplines. In chemistry, for example, carbon atoms are black, hydrogen is white, oxygen is red, and nitrogen is blue. Illustrations involving these elements, of course, would have to be composed around the prespecified colors.

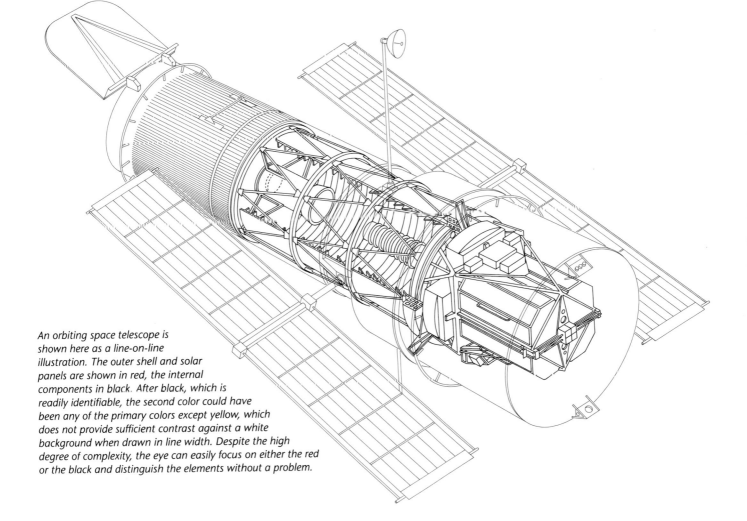

An orbiting space telescope is shown here as a line-on-line illustration. The outer shell and solar panels are shown in red, the internal components in black. After black, which is readily identifiable, the second color could have been any of the primary colors except yellow, which does not provide sufficient contrast against a white background when drawn in line width. Despite the high degree of complexity, the eye can easily focus on either the red or the black and distinguish the elements without a problem.

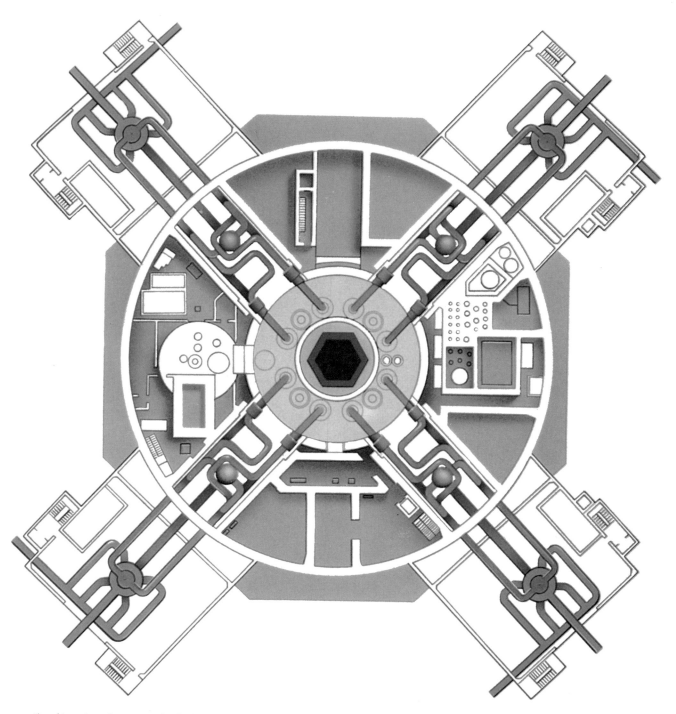

This cross-sectional top view of a reactor clearly displays all the support structure and equipment, but here these elements are reduced to subordinate roles. The strong color against the neutral gray background highlights the reactor core and the circuitry of the systems emanating from it. The use of black line, gray, white, and color gives the illustration clarity and affords a quick "read-out" of information. Working white into the illustration helps to separate the components and makes for an elegant graphic design.

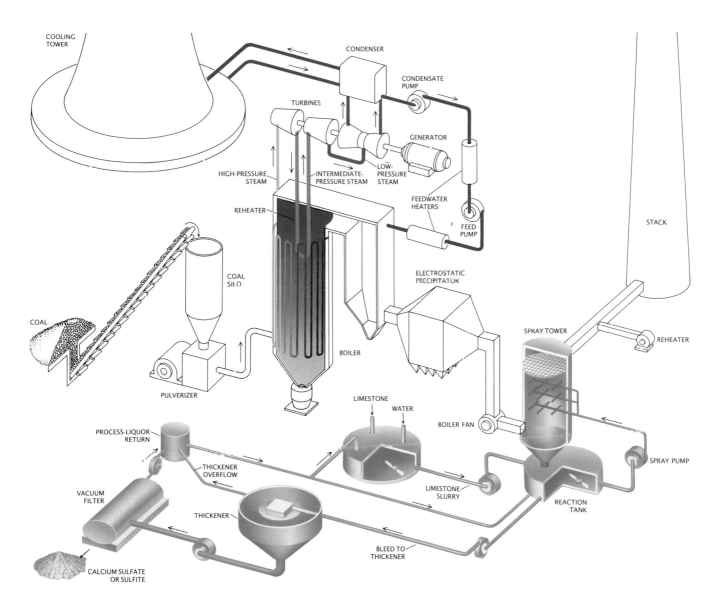

COOLING TOWER

CONDENSER

CONDENSATE PUMP

TURBINES

GENERATOR

HIGH-PRESSURE STEAM

INTERMEDIATE-PRESSURE STEAM

LOW-PRESSURE STEAM

REHEATER

FEEDWATER HEATERS

F FEED PUMP

STACK

COAL SILO

ELECTROSTATIC PRECIPITATOR

SPRAY TOWER

REHEATER

COAL

BOILER

PULVERIZER

LIMESTONE

WATER

BOILER FAN

SPRAY PUMP

PROCESS-LIQUOR RETURN

THICKENER OVERFLOW

LIMESTONE SLURRY

REACTION TANK

VACUUM FILTER

THICKENER

BLEED TO THICKENER

CALCIUM SULFATE OR SULFITE

In this complicated industrial-process flow diagram, color is used to identify systems described in the text. Where there are several systems to be identified, the colors chosen should be clear and distinct. Since the colors here served no purpose beyond distinguishing the various systems, any pleasing combination could have been used.

CUTTING THROUGH CLUTTER

While a fully integrated mechanical structure is needed to operate a piece of machinery properly, it is not always necessary to see all of the parts in order to understand how the machine works. I would not need to show all of the cables on a suspension bridge, for example—a well-spaced representative number of them would be adequate for the reader to comprehend the principle of suspension. Moreover, showing every last component often distracts the eye or obscures important details. Successful illustrations often depend on knowing what to leave in and what can be left out.

Spatial distortions are sometimes necessary as well. This too is acceptable, as long as it does not render the object unrecognizable. In general, these liberties can be taken at the discretion of the artist, but should be discussed with the author, editor, and art director.

These two pieces of sophisticated equipment were analyzed and redesigned to enable the reader to view the internal structures and follow the mechanical operations taking place within them. Trim and component parts not essential to the design objective were eliminated.

For this illustration of a ball control for a video game, most of the internal support structures were eliminated to reveal the mechanical and electronic parts. A cutaway approach would have been less effective here, since the cutaway section's jagged edges would have imposed an extraneous configuration into a relatively small area, thereby complicating the situation instead of simplifying it. Instead, the walls of the plastic case were rendered translucent. The white control ball in the center was treated in the same manner. The detail in the lower-left corner is an enlarged view of the light-beam assembly, which receives pulses from the perforated black discs and, in turn, signals the electronic device that controls the action of the video screen.

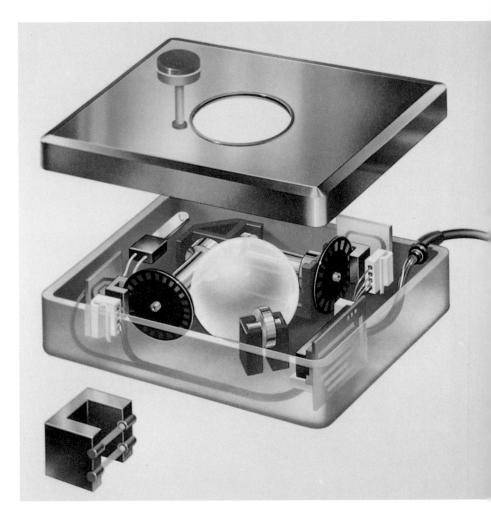

This machine, used in the treatment of blood diseases, irradiates a thin film of blood with ultraviolet rays as the blood passes through a clear plastic frame. The actual width of the machine is 16 inches (41 centimeters), but the full-color rendering of the machine had to be drawn twice as wide in order to show all the components clearly. Although the twin banks of ultraviolet lamps in the actual device almost touched the plastic frame, I increased the proportions of the machine and pulled the lamps away from the frame to show the entire flow process. The diagrammatic flow diagram on the left shows the entire procedure.

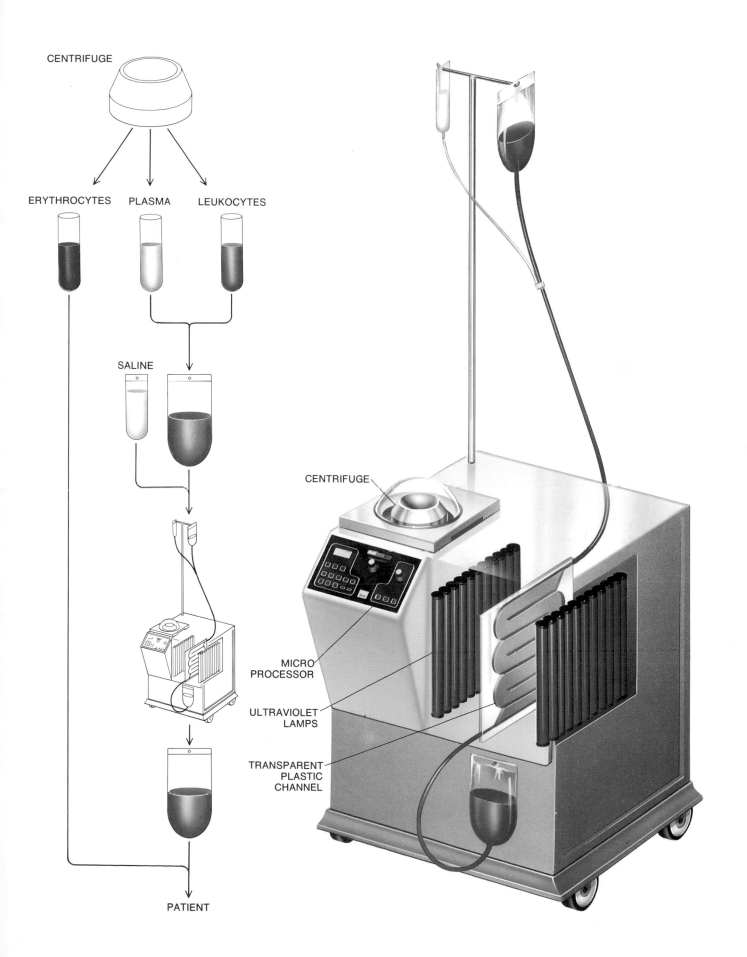

CENTRIFUGE

ERYTHROCYTES   PLASMA   LEUKOCYTES

SALINE

CENTRIFUGE

MICRO
PROCESSOR

ULTRAVIOLET
LAMPS

TRANSPARENT
PLASTIC
CHANNEL

PATIENT

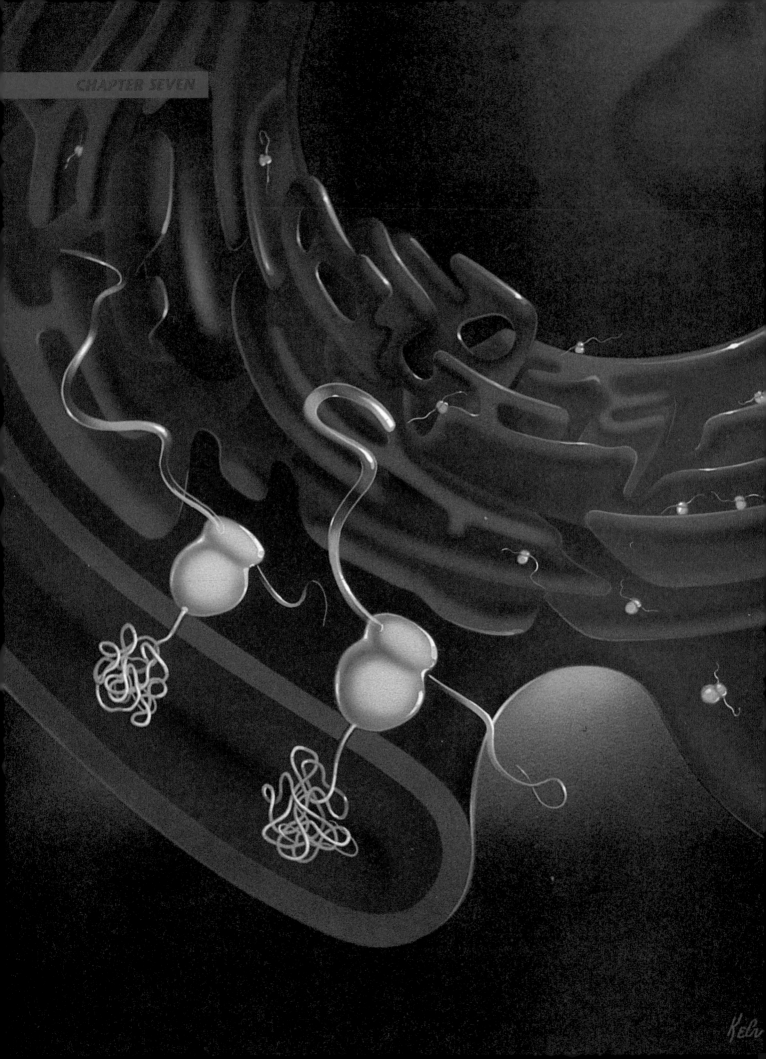

CHAPTER SEVEN

MEDICAL AND BIOLOGICAL ILLUSTRATION

Medical and biological illustration covers a range of practical functions. Illustrating for medical textbooks and for diagnostic and treatment purposes requires such an imposing range of knowledge and technical skill that it effectively necessitates medical training as well as an art background. This level of illustration is essentially a specialty unto itself, and many of its foremost practitioners have actually attended medical school.

Illustration for medical magazines and for pharmaceutical and advertising companies, while still technically demanding, places more emphasis on design. This category of medical illustration, while requiring a solid foundation in anatomy and drawing techniques, does not require an imposing biology-intensive background, and is more compatible with the training of the typical scientific illustrator.

In addition, there is a large body of reliable reference material available. Art students may find Fritz Schider's *An Atlas of Anatomy for Artists* (Dover) a good basic drawing guide. On a more professional level, Frank N. Netter's *The CIBA Collection of Medical Illustrations* (CIBA Medical Education Division) and Peter Williams and Roger Warwick's *Gray's Anatomy* (Saunders) are also invaluable. In addition, scientific suppliers sell anatomically correct breakaway models showing the relationships of organs to muscles, and so on. Detailed anatomical charts are available as well, and these make useful drawing guides.

The images included here point to some of the ways color and design can be used to defuse potentially unsettling aspects of the subject matter while simultaneously transforming it into dramatic, attractive imagery.

This highly magnified segment of a cell shows where protein is produced. Since cells have little or no color, I was not limited to a prescribed palette and was therefore free to choose colors for their overall effectiveness. The cell nucleus is shown in blue, with the mitochondrion in green, the ribosomes in yellow, and the endoplasmic reticulum in red.

FINDING SPECIFIC DETAIL WITHIN BROAD PATTERN

Through the use of creative imagery, an abundance of information can be packaged into a single illustration and still permit a wide range of design opportunities.

Starting with some recognizable background view—commonly in the form of a silhouette, a realistic view, or a schematic representation—and then superimposing an enlarged section against it showcases the significant detail as well as the environment from which it was derived.

This juxtaposition of hand and cell shows the effects of arthritis, while the floating pills and tablets suggest possible drug therapies. The presence of the human cell creates a unifying element for the otherwise disassociated elements. By giving a luminous quality to the cell and superimposing the x-ray of the hand, the point of the subject matter is easily conveyed.

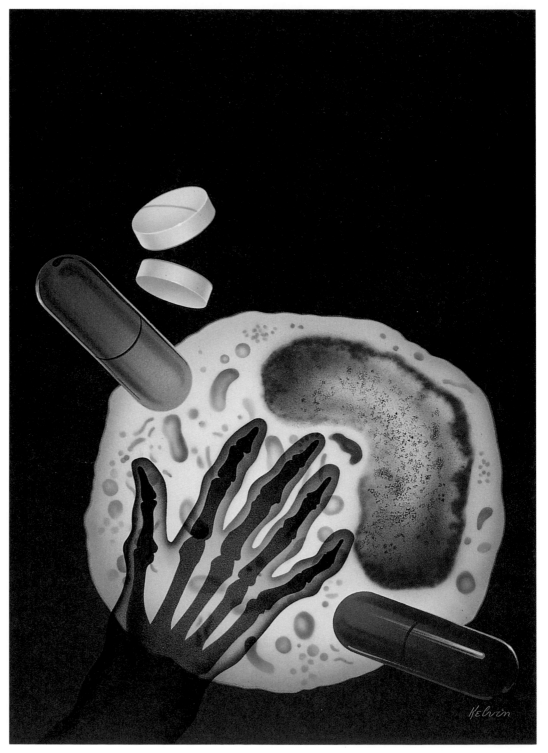

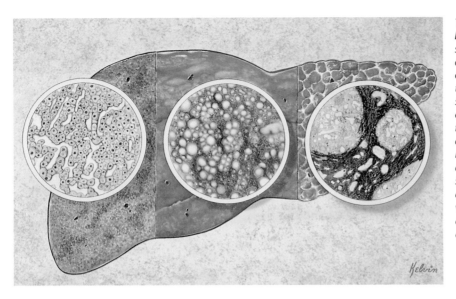

This illustration was originally planned as a six-box design—three boxes on top showing cross sections of the liver in progressive stages of the disease and three below with microscopic views of the cellular structure. However, achieving the necessary level of detail while simultaneously fitting all six boxes within the cover design's space restrictions would have forced the elements to be too small to be of any value. The final concept shows a single recognizable cross section of a human liver divided into three segments, each rendered to show the progression of the disease. Floating above each section is its microscopic view. This version, in addition to having a more pleasing design than the first concept, transmits the crucial information more readily.

This illustration depicts the transmission of an impulse from a neuron to a dendrite via a chemical neurotransmitter called choline. The translucent treatment of the left-foreground neuron (in red) and the right-foreground dendrite (in blue) permits the stylized rendition of the brain in the background to remain totally visible.

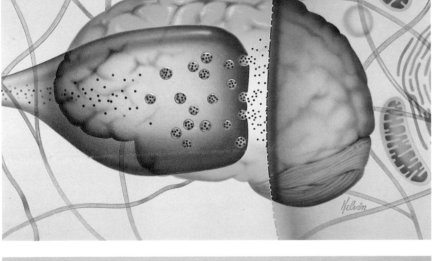

To highlight the medulla, the center portion of the adrenal gland located on the kidney, I positioned it in a rectangle above the enlargement of its cellular structure. A light airbrush spray of zinc white over the background cells gives it a hazy, translucent look.

USING LIGHT DRAMATICALLY

Just as lighting is used to produce special effects on the stage of a theater, it can also be used to achieve various effects in a painting. The simplest way to add a touch of drama and isolate various components is by backlighting the subject from a single light source. Using colored light from multiple rear sources, or from a combination of front and rear, helps to focus attention on several defined areas without fragmenting the subject. Exciting visual effects can also be achieved by using one or more of the component elements as the illumination source for the surrounding objects.

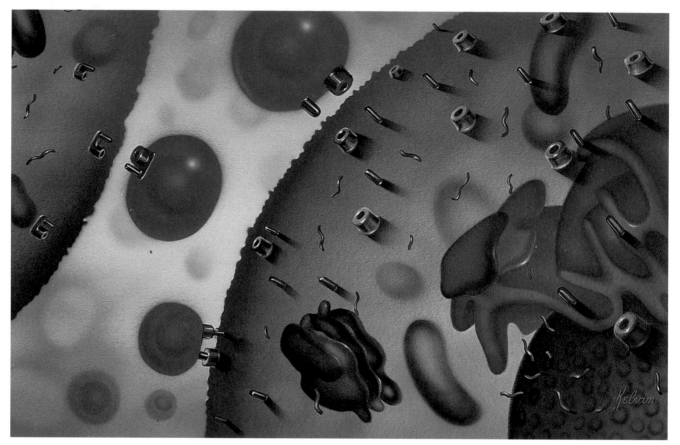

The representation of the action in this view of cells, antibodies, and antigens in a Star Wars–like battle for survival is partially realistic, partially schematic. Dramatic backlighting silhouettes and defines the cells.

For an article on diabetes, I rendered this undulating wall of cells as though it were a landscape. Unusual illumination presents a unique departure from the traditional treatment of cellular structures, and the lighting transition from warm reds and yellow in the foreground to cool blues in the background accentuates the rolling effect.

TONING DOWN SEVERE SUBJECT MATTER

When intimidating subject matter is prepared for a nonprofessional audience, moderation is often called for. For the illustrations shown here, designed for the general market reached by the Time Inc. Science Library Series, I intentionally avoided the customary way of presenting medical material. Given the strong instinctive reactions many people have to images of blood, internal organs, and so forth, a stylized treatment of the body parts seemed a much more suitable approach than the literal depiction of organs and tissues.

The simple, white silhouette of the body and the pale, muted colors of the organs serve as a soft background for the metallic form of the artificial heart and power supply, thereby presenting an unpleasant concept in a nonthreatening manner. Since specific colors are known to trigger different emotional responses, the color palette was of particular importance. Soft background colors, such as those used here, tend to have a calming effect, whereas darker, stronger colors probably would have appeared ominous in this context.

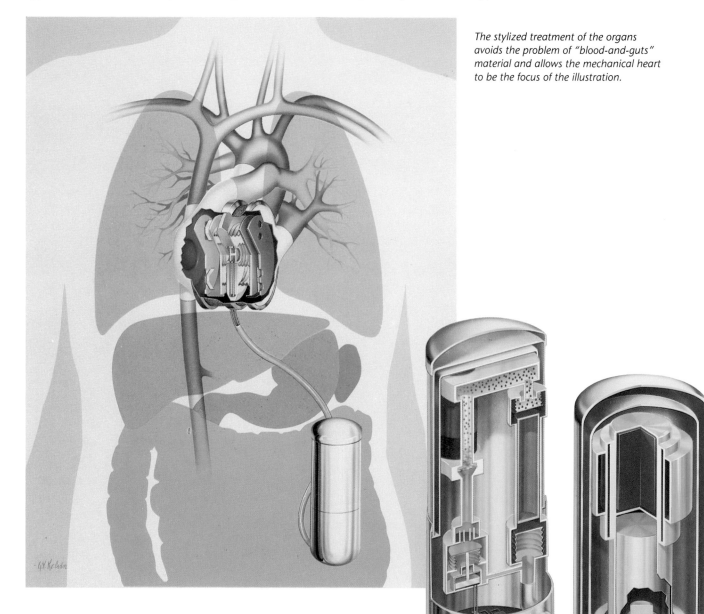

The stylized treatment of the organs avoids the problem of "blood-and-guts" material and allows the mechanical heart to be the focus of the illustration.

These cutaway/cross-sectional views reveal the internal workings of the artificial heart power sources.

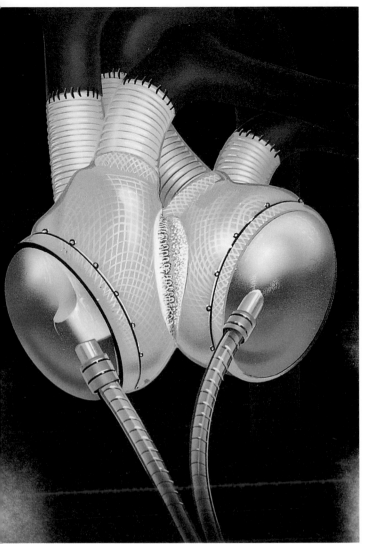

This portrait of a brightly illuminated mechanical heart is presented against a simple, dark background.

Here a mechanical heart appears against the shadowy image of the rib cage. I left an acetate mask of the rib cage in place while I laid down the background color; halfway through the rendering, the mask was removed and the rendering was continued to completion. The lighter, x-ray appearance of the ribs was due to reduced pigment application to that area when the mask was in place. Working with different densities of a single pigment in this manner produces a truer color than can be achieved by spraying a thin coat of white over a background color.

Magnet

Hall effect sensor

THE EVOLVING PICTURE OF THE AIDS VIRUS

Can a deadly virus be seen as a thing of beauty? Little did I know just how important these portraits of the AIDS virus would turn out to be. The first painting appeared on the cover of *Scientific American* in 1987, only three years after the virus itself had been identified.

Under the guidance of the leading experts in the field, the first painting was prepared using the research data that was available at that time. Information regarding shape, size, and structure had been gathered by molecular biologists using electron microscopy and x-ray diffraction patterns.

The illustration, showing the virus at one million times its actual size, provoked tremendous reaction. Once published, requests for permission to reproduce the art came in from all over. It had become the world's first definitive portrait of the AIDS virus.

Chet Raymo, science reporter for the *Boston Globe*, wrote:

I have returned to the cover illustration again and again. It is a simplified rendering of the overall architecture of the virus. One might expect that the agent of so pernicious a disease would itself be unlovely. But it is a thing of spare elegance, with the delicacy of a flower and the symmetry of a snowflake. It has no foul aspect. It is starkly and simply beautiful.

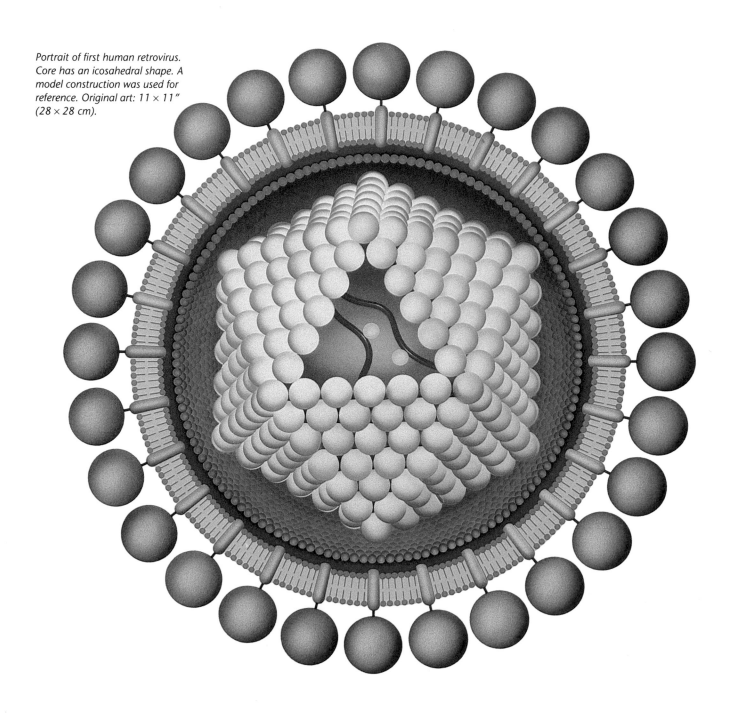

Portrait of first human retrovirus. Core has an icosahedral shape. A model construction was used for reference. Original art: 11 × 11" (28 × 28 cm).

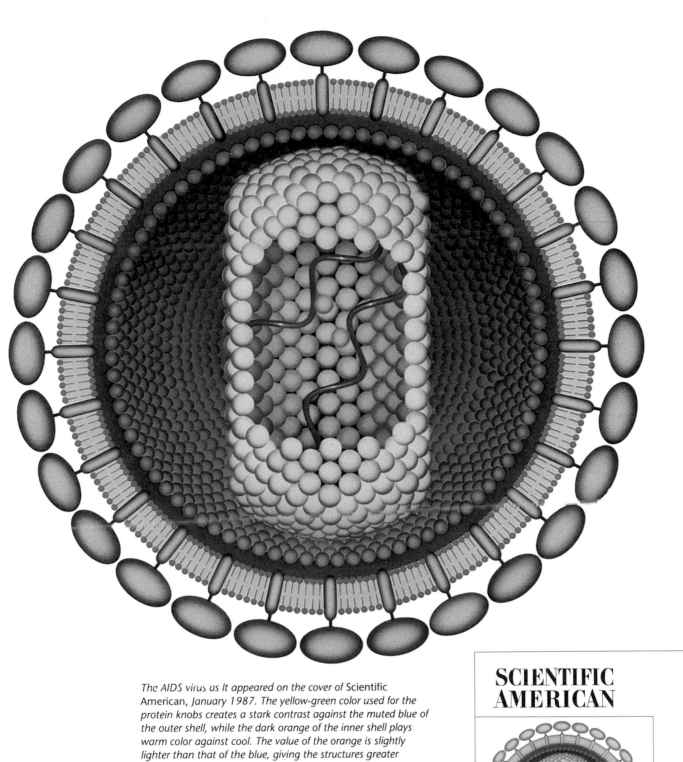

The AIDS virus as It appeared on the cover of Scientific American, January 1987. The yellow-green color used for the protein knobs creates a stark contrast against the muted blue of the outer shell, while the dark orange of the inner shell plays warm color against cool. The value of the orange is slightly lighter than that of the blue, giving the structures greater separation. The lightest color, pale ivory, was used for the core in order to focus attention on the RNA helix, in red, at its center. With the exception of the outer knobs, all of the colors are muted, allowing the geometry of the virus to carry the design. Original art: 11 × 11" (28 × 28 cm).

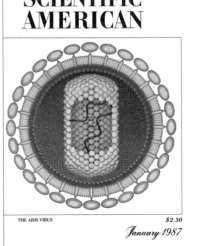

Within a short time, intensified research led to more definitive data on the organism. By October of 1988, a more precise schematic portrait appeared in *Scientific American*, this time with fewer protein knobs on the outer shell, the RNA covered by protein, and the core structure rendered in its newly determined conical shape. Because color is not present at the molecular level, I was left to my own artistic discretion in choosing colors to represent the different components. As you can see, the paintings were done with two completely different color palettes, so as to readily distinguish the two illustrations from each other.

For the revised portrait, I was given new dimensional and spacing information to work from. I divided a circle into 18 equal parts to correctly space the outside protein knobs and stalks, which are distinguishing characteristics of the virus. I made a pie-shaped acetate template of one section of the circle and, placing a pin at the center, used step-and-repeat indexing to draw the orange protein knobs, the brown balls of the outer lipid bilayer, and the blue balls on the rim of the inner shell. Indexing was used again to do the airbrush work for the final painting, and also to render the remaining molecular structure.

The S-shaped RNA structure, seen in yellow in the center of the core, required a different approach. Although there were no pictures available, it was now known that this structure was covered with protein molecules. Working from a verbal description, I determined that these could be schematically represented as linked spheres. A simple model was constructed by wrapping a long string of beads around a length of rubber tubing and then taping the ends of the string to the ends of the tubing. Bending the tube into an S curve produced a reasonable simulation of what protein-covered RNA should look like. From a double-enlarged close-up photo of the beads, I traced off the spiraling shapes, giving me the right effect and perspective.

These AIDS virus illustrations reinforce a point I have made elsewhere: Scientific and technical changes and advances occur so rapidly today that almost no image can ever be considered as final—the business of revising and updating illustrations is an ongoing process. Even as this book goes to press, updated research material is being gathered from all parts of the world. We can only hope that the next portrait of the virus points in the direction of a cure for the insidious disease it represents.

Indexing template used to create "step-and-repeat" pattern for the pencil layout and final airbrush rendering.

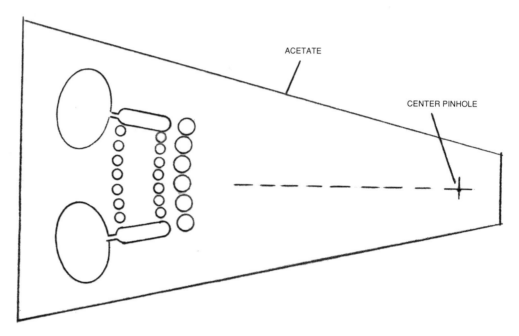

ACETATE

CENTER PINHOLE

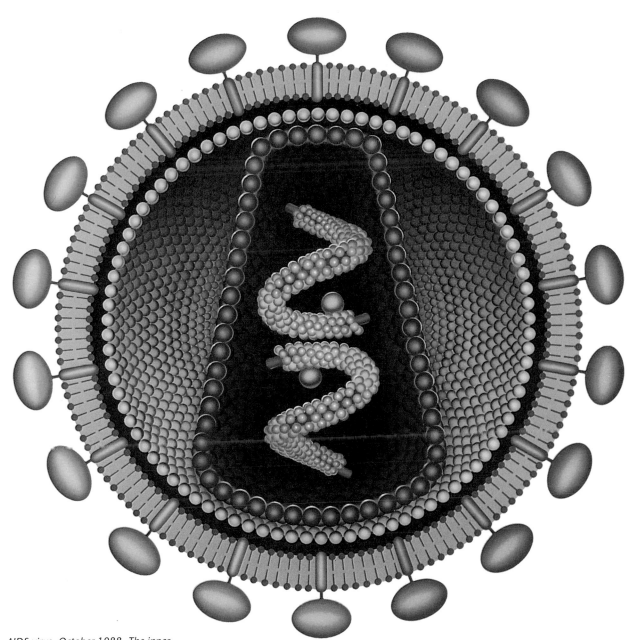

AIDS virus, October 1988. The inner structure of the tapered core has been darkened, allowing the ball shapes to fade toward the back and center without losing their identifiable shapes. The protein-covered RNA, in yellow, shows up well against the diffused dark-red pattern. The brighter blue of the inner shell clearly differentiates the structures. Original art: 11 × 11" (28 × 28 cm).

MAPPING OUT THE STORY OF DNA

The 1953 discovery of the DNA structure (deoxyribonucleic acid) by James A. Watson and Francis Crick is destined to change the very course of human development in the coming century. Indeed, in just the few short years since the discovery, we have already witnessed astonishing breakthroughs in genetic engineering.

DNA, a fascinating subject to depict, has unique identifying visual characteristics: It is a right-handed double helix (that is, it has a clockwise rotation) containing a major and minor groove with 10 base pairs of amino acids in every complete turn. It is usually presented in a stylized schematic form. Having first depicted DNA in 1962, I later had the privilege of illustrating Dr. Watson's book, *The DNA Story*, and have since produced several hundred DNA illustrations for textbooks, magazines, and media of every type.

The illustrations shown here, designed in consultation with Dr. Watson, tell part of the story of DNA. In addition they feature just about all of the airbrush, model building, and photographic techniques discussed over the course of this book. Note the soft, airbrushed edge on the nucleus and the transition from hard to soft edges on the lipid granules. The centers of the mitochondrion were airbrushed, while the fingerlike structures were painted in by hand. The white boxes seem to be floating above the painting, owing to soft shadows behind the boxes, which create a three-dimensional effect.

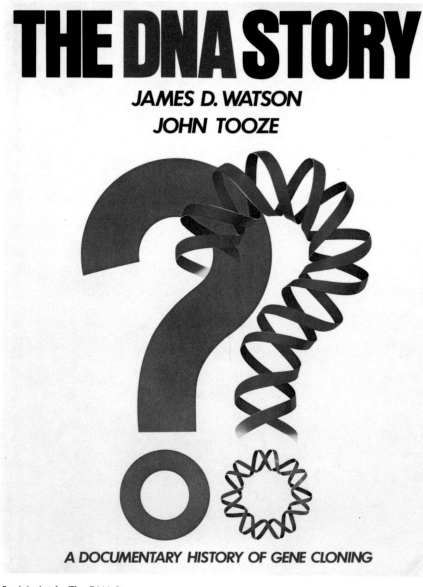

Book jacket for The DNA Story.

An enlarged section of a typical cell introduces DNA in its environment. Photos taken with an electron microscope served as reference. A drop-out segment of a chromosome in the cell nucleus, shown as a small, white box, locates the DNA, and the arrow directs the eye to an enlargement of the chromosome, showing the DNA as coiled, threadlike structures. A second arrow directs attention to a schematized representation of the DNA's molecular structure.

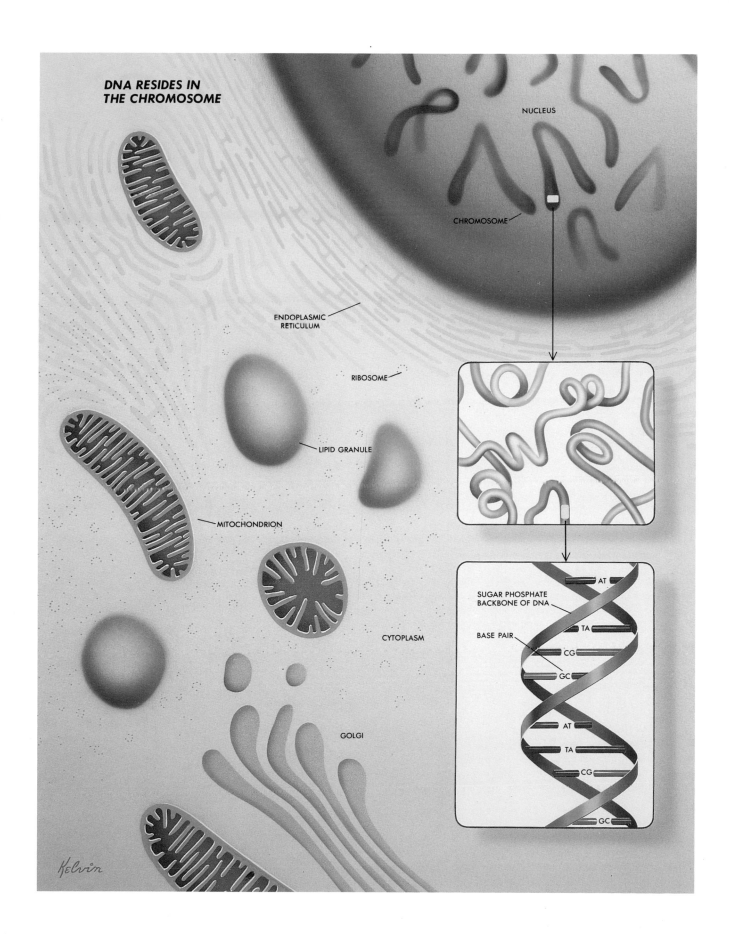

DNA RESIDES IN THE CHROMOSOME

NUCLEUS

CHROMOSOME

ENDOPLASMIC RETICULUM

RIBOSOME

LIPID GRANULE

MITOCHONDRION

CYTOPLASM

GOLGI

Kelvin

SUGAR PHOSPHATE BACKBONE OF DNA

BASE PAIR

AT

TA

CG

GC

AT

TA

CG

GC

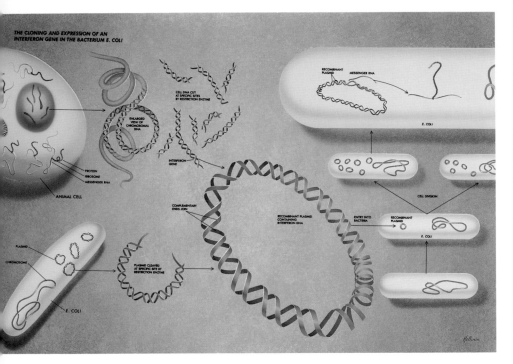

THE CLONING AND EXPRESSION OF AN
INTERFERON GENE IN THE BACTERIUM E. COLI

CELL DNA CUT
AT SPECIFIC SITES
BY RESTRICTION ENZYME

RECOMBINANT
PLASMID

MESSENGER RNA

ENLARGED
VIEW OF
CHROMOSOMAL
DNA

E. COLI

INTERFERON
GENE

PROTEIN
RIBOSOME
MESSENGER RNA

ANIMAL CELL

COMPLEMENTARY
ENDS JOIN

CELL DIVISION

PLASMID

CHROMOSOME

RECOMBINANT PLASMID
CONTAINING
INTERFERON DNA

ENTRY INTO
BACTERIA

RECOMBINANT
PLASMID

E. COLI

E. COLI

PLASMID CLEAVED
AT SPECIFIC SITE BY
RESTRICTION ENZYME

E. COLI

INFORMATION FLOWS FROM DNA THROUGH

INSIDE THE
CELL NUCLEUS

Once again, the story is conveyed with arrows to direct the reader. This image maintains continuity with the previous illustration by using the same colors. In this case, the background is not intrinsic to the story, and so is used instead to highlight different components and create interest.

The softly textured background was accomplished by daubing the surface alternately with two sponges—one saturated with a watery mixture of turquoise blue and indigo, the other with zinc white. A fine spray of turquoise blue was used to deepen the color at both ends of the background, creating a highlighted center area.

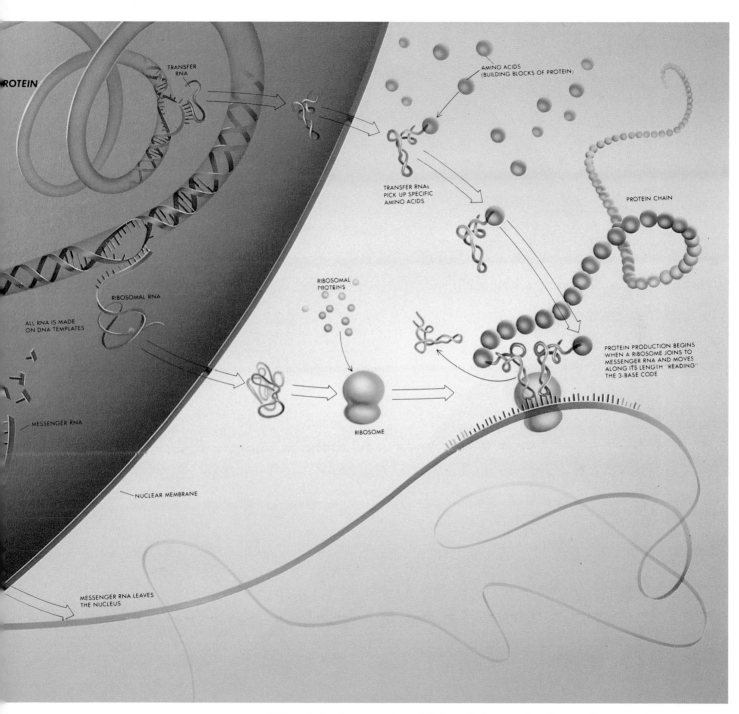

PROTEIN

TRANSFER
RNA

AMINO ACIDS
(BUILDING BLOCKS OF PROTEIN)

TRANSFER RNAs
PICK UP SPECIFIC
AMINO ACIDS

PROTEIN CHAIN

RIBOSOMAL
PROTEINS

RIBOSOMAL RNA

ALL RNA IS MADE
ON DNA TEMPLATES

PROTEIN PRODUCTION BEGINS
WHEN A RIBOSOME JOINS TO
MESSENGER RNA AND MOVES
ALONG ITS LENGTH "READING"
THE 3-BASE CODE

MESSENGER RNA

RIBOSOME

NUCLEAR MEMBRANE

MESSENGER RNA LEAVES
THE NUCLEUS

Still employing the same color and design techniques used on the preceding pages, this schematic representation of the DNA double helix is subtly faded in at the point where the information sequence begins. The DNA shown fading out in the background suggests that the story continues beyond the limits of this illustration.

Note that a new arrow style has been introduced to differentiate primary and secondary paths. Arrows can be strong graphic devices but require careful planning regarding size, shape, and color, or they can easily overpower the subject matter. These arrows, for example, though large in size, feature a simple line design that does not dominate the illustration.

CHAPTER EIGHT

GEOGRAPHIC AND GEOLOGICAL SUBJECTS

It should be clear by now that a scientific illustrator must have a reasonable level of competence in many areas. The previous chapter examined some of the concerns and possibilities pertaining to depicting the relatively small confines of the human body and much smaller (even microscopic) biological subjects; this chapter shifts the emphasis from micro to macro, exploring the illustration of the Earth itself, including its inner geological workings and its outer land masses.

In one way or another, much of this work involves maps. However, the scientific illustrator is not a cartographer—cartography is an art in itself. Nevertheless, a good many map assignments have crossed my desk over the years. In most cases, they have entailed selecting a reference map that most closely fills the needs of the assignment and tracing off the pertinent information. Happily, this is one time when the artist does not have to "reinvent the wheel," so to speak, as every type of map projection can be found in any map store, and atlases are excellent reference sources as well. The artist's task is to adapt the map to fit the needs of the assignment with tints, textures, or any of the other design stylings that would serve to enhance the project.

Similarly, color yields particularly rewarding results for geological illustration, a dynamic subject area, full of creative possibilities, that has been saddled with drab line art for far too long.

A view of the Earth directly in front of the sun, with solar flares visible. The Earth was represented by a globe, which was photographed with high-speed, fine-grain film. The photo was then enlarged and traced. The side spraying technique (see Chapter 6) was used to achieve texture.

MAKING MAPS USING LINE, TINT, AND TEXTURE

The two maps of Africa shown here, both prepared for a high school textbook, have been given two different treatments in order to accommodate diverse needs. The first map was handled as a simple line drawing derived by tracing over a basic reference map of the continent. Since only border lines were required, all other map features were excluded. The second map was designed to be decorative as well as more informative. In this instance, side spraying the surface, which was first built up with white acrylic paint, lends an interesting textured pattern to the terrain surrounding the lake areas. Had it been necessary to suggest additional features, such as mountains and forests, a variety of textures could have been employed.

Flat tints work nicely for suggesting motion or division of space, as in the stylized map on page 127 depicting worldwide oceanic currents. The flow of color defines the direction of the moving currents, creating a pleasing design element as well.

For a map that must show multiple features, several techniques may be combined, as was done for the map of Europe at the bottom of page 127. The ancient sites of the Mithraic temples are easily seen against the flat-tinted background, while airbrush painting defines the continental borders.

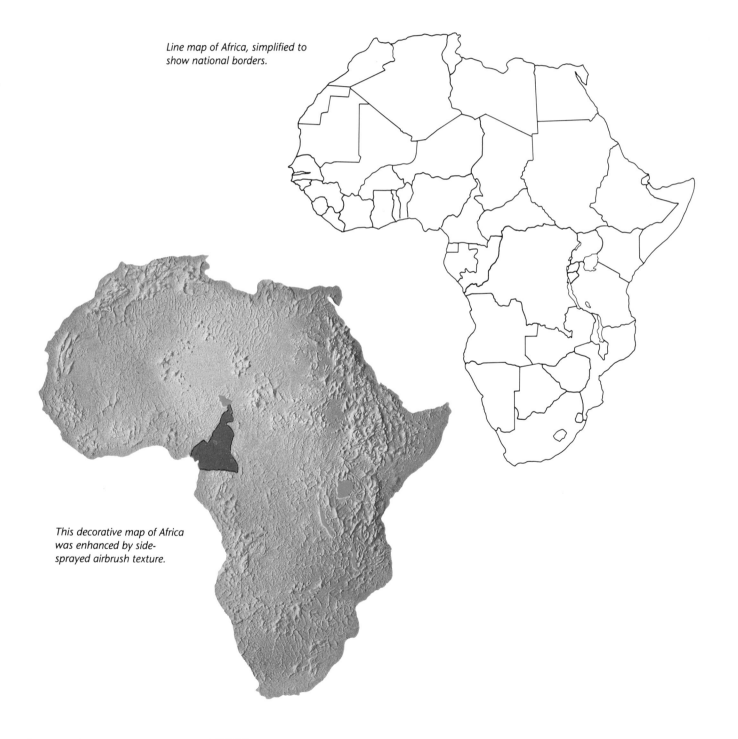

Line map of Africa, simplified to show national borders.

This decorative map of Africa was enhanced by side-sprayed airbrush texture.

This stylized map uses flat tints to indicate oceanic currents.

Flat tints enhanced by airbrushing create a deeper tone around the continental borders. The dots show the location of Mithraic temples during the time of the Roman Empire.

CREATING A THREE-DIMENSIONAL RELIEF MAP

This distinctive black-and-white relief map is more than just a simple illustration—it is the result of a combination of graphics, model making, and photography. It also has an interesting background story.

The map, prepared for a book in a series of Time-Life nature and science books, was not originally assigned to me. It was given to a fellow illustrator, who planned to execute it as a traditional oil painting. As his deadline neared, it became clear that he would not complete the assignment in time. With the book already laid out and a page and half allocated for the map, I was called in on the slim chance that I could design, paint, and deliver something suitable in two days.

To save time, I glued the original artist's tissue layout to a piece of two-ply illustration board and cut out the continental shapes with a jigsaw. Then I covered the entire surface with gesso—a thick, white primer—which I picked at and molded until the shapes took on the right design and proportion. Using dark-gray paint, I employed low-angle side spraying to form natural-looking shadows on the textured terrain.

At this stage, the map could have been considered finished, but I had a plan to take it one step further: The component shapes were arranged on 1-inch (2.5-centimeter) pins above a background board, illuminated from the side, and photographed. The resulting picture is seen here.

Raising the continents above the oceans and providing illumination from one side produced a dramatic play of shadows.

RADIO-WAVE MAPS

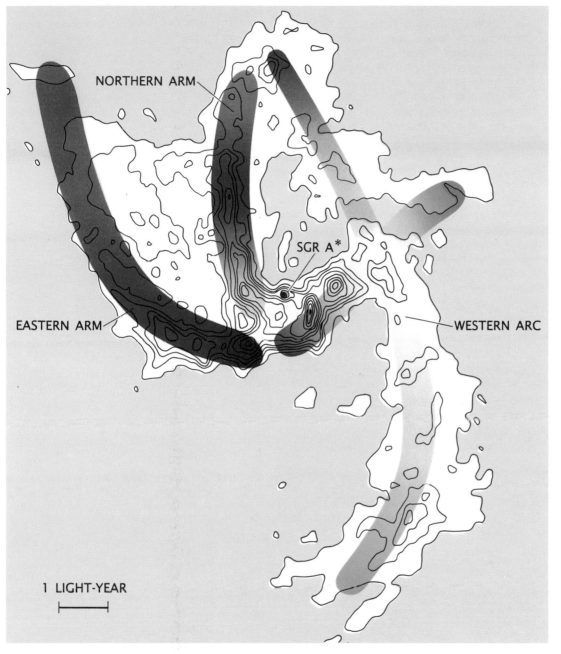

NORTHERN ARM

SGR A*

EASTERN ARM

WESTERN ARC

1 LIGHT-YEAR

Although this map showing the intensity of radio waves in a distant galaxy appears somewhat stylized, it is actually a tracing. The basic shapes were traced from a computer-generated map furnished by the author; color was added to make data easy to identify.

Radio-wave maps may be unfamiliar to the general public but are commonly used in astronomy. Radio telescopes detect energy emissions and the configurations of their force from distant galaxies, which are recorded by computers and displayed in the form of maps, similar to topographical maps. Completely computer-generated, the graphs usually are not suitable for reproduction and have to be traced in black line.

EXTRAPOLATING THE THIRD DIMENSION FROM A FLAT MAP

This illustration for a book entitled *Flight*, part of the Time Inc. Science Library Series, diagrams all the commercial air routes between O'Hare Airport in Chicago and Kennedy Airport in New York. When designing it, I wanted to convey to the reader the same sense of three-dimensional space that pilots employ when flying their planes.

The initial editorial concept was to show an air-route map superimposed against a background map of the northern United States. After some discussion, I proposed what I believed to be a more imaginative way of handling the project, one that called for showing the air routes as they really are—a multilayered network of highways and superhighways crisscrossing the skies. I could already see it in my mind: a three-dimensional view of space, with stacks of heretofore invisible highways. And by making each highway level translucent, it would be possible to see all the network's details.

The editor provided me with an air-route map (below), which I superimposed onto a standard map of the United States. After first selecting a good viewing angle of the three important control centers referred to in the text—Kennedy, O'Hare and St. Louis—I photographed the combined maps, which enabled me to convert the orthographic plan into a perspective one, complete with horizon line and vanishing points. This became the base from which I projected the air routes. The vertical scale had to be exaggerated substantially, since there are approximately 700 miles between airports and the highest commercial jet route is only about 6 miles up.

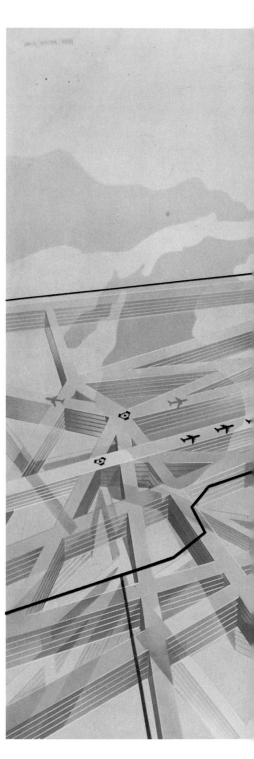

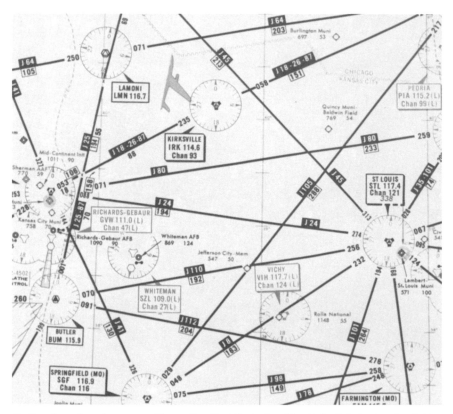

Portion of an air-route map used in creating final illustration. The St. Louis hub at center right corresponds to the air-route intersection point seen at center right of the finished illustration.

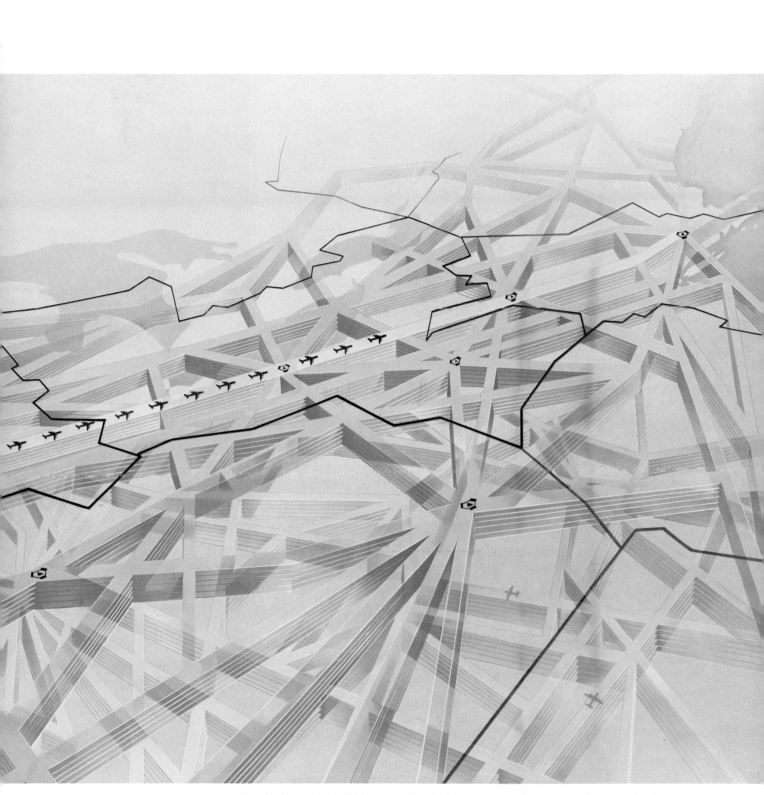

The color-keying of the flight lanes—yellow for jet planes, green for propeller planes—makes the routes easier to follow and contributes to the design's overall impact. Original art: 16 × 27" (41 × 69 cm).

PLOTTING DATA ONTO MAP PROJECTIONS

This cover painting for *Scientific American* (the magazine's first use of a foldout cover) makes a powerful presentation of humankind's impact on our planet. Each point of light represents a major expenditure of energy taking place somewhere on Earth.

This assignment required the use of a *map projection* to represent the three-dimensional Earth on a flat, two-dimensional plane. There are many types of flat projections, but none of them represents the Earth's spherical surface without some degree of distortion. For illustration purposes, one of the most commonly used maps is the Mercator projection, which presents the Earth's surface in a rectangular plane; but since I wanted a curved horizon line, I chose the oval-shaped Robinson projection.

The locations of the light points are not arbitrary. Much of the reference for them came from satellite photographs taken over a 10-year period. Energy source activity patterns, such as those caused by natural gas flares, city lights, and slash-burning of forests, were systematically photographed and plotted onto a Mercator map projection. Using an atlas as my cross-reference, I replotted each point in accordance with its longitude and latitude onto the Robinson projection.

The working size was 18 by 30 inches (46 by 76 centimeters) on four-ply bainbridge board. The entire board was sprayed ivory black, followed by a spray of a diluted gum arabic mixture to stabilize the surface. Acetate masks of the land areas were cut and positioned with low-tack adhesive. The curved mask for the horizon was scored using a compass and a four-foot (1.2-meter) straightedge fitted with a needle on one end.

Once all the masks were in place, I airbrushed in the graduated blue tone of the oceans. I wanted the horizon line to have that glowing effect and soft edge so often seen in satellite photos. To achieve this, I went back with white and a lighter blue on the horizon line, moving the mask slightly after each pass with the airbrush to produce the glow. A fine spray of turquoise blue and some white highlighting brought a glow to the continents. With this much completed, I removed the masks covering the land areas, which had remained the flat black of the background.

Next I had to transfer the points of light accurately from my pencil layout to the illustration. Putting a dot of light-gray pastel on the reverse side of the layout directly beneath each light point, I used pinhole guides to position the layout in perfect register on top of the illustration. Slight pressure from a pencil point transferred the gray dots onto the illustration. I then went over each gray dot freehand with a wrico pen filled with pure white paint, followed by an airbrushed spray of lemon yellow. The yellow sprayed over the white produced the highly reflective effect, and the overspray produced the luminous glow surrounding the points. A light spray of zinc white over the flat black land areas completed the painting.

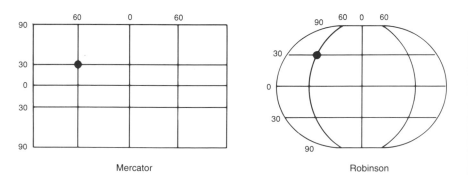

Mercator and Robinson projections, each showing a point located at 30 degrees north latitude and 60 degrees west longitude. Although the two projection systems differ, the latitudinal and longitudinal values they show are the same, allowing points on one system to be transferred to the other.

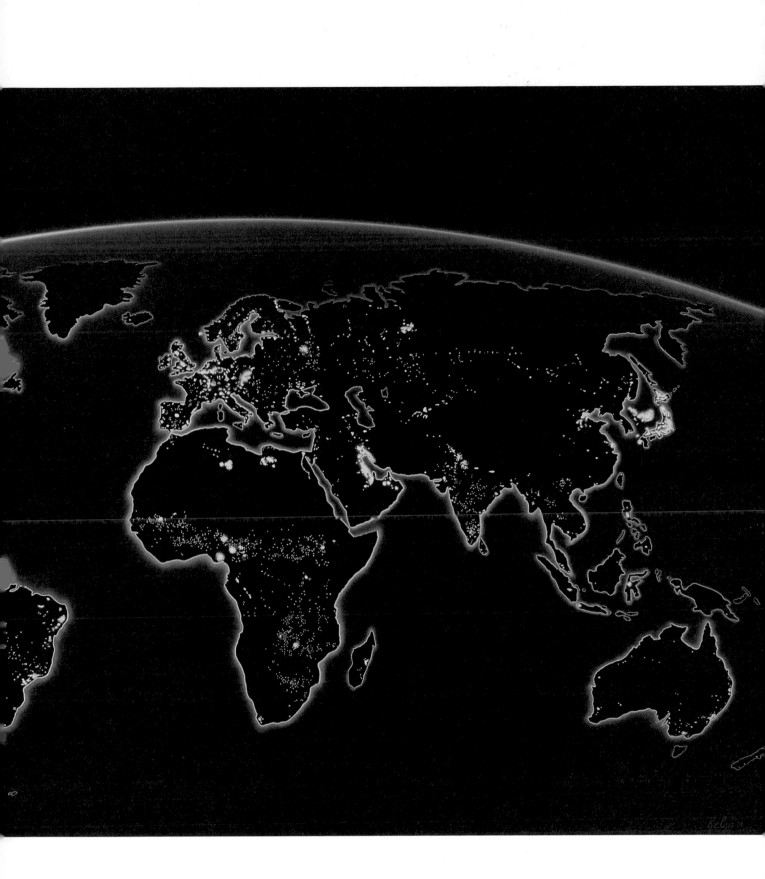

REPRESENTING GEOLOGICAL FEATURES REALISTICALLY

I make every effort to look at each assignment with a fresh eye, trying to avoid routine treatments and looking for new approaches to the material. Unfortunately, as must be apparent to anyone flipping through the pages of the numerous geology books published in recent years, not all illustrators make such an effort. Why has geology, a wonderful subject so rich in textures, shapes, and colors, been relegated to a visual world inhabited almost exclusively by black line illustrations? Although the subject is fascinating, geology textbooks have not changed substantially in decades.

While an interesting array of textures can be used for geology, most texts only present a variety of cross-hatching and stippling patterns to represent the different strata of Earth and rock. Some recent books have tried to be a bit more creative by adding flat color tints and watercolor washes; while this is an inexpensive way to introduce color into the text, it gives the illustrations a cartoonlike look that tends to trivialize the subject matter.

I had the opportunity to introduce a realistic approach to geological illustration while preparing samples for *Modern Physical Geology*, a college-level textbook published by Saunders College Publishing in 1991. I proposed a complete departure from the prevailing style in the field—instead of the old-fashioned black line and flat tint illustrations, we would go with full-color, three-dimensional drawings, giving students a greater sense of form, configuration and structure. These illustrations were specifically designed to make it easy for the students to identify formations falling within their own frames of reference, and to follow and identify sequential geological processes.

All the colors and textures used in the book were first set down in a reference key, thereby establishing identifiable characteristics that would be easily recognized when they appeared elsewhere in the book. Sandstone, for example, would appear with the same color and texture throughout, regardless of the context.

I also was able to address a problem I had seen repeated many times. Geology books often feature *carpenter's projections*, a type of false, distorted perspective drawing often used by cabinetmakers that presents a front view of an object with a side plane drawn off at an angle. Instead, I planned all the illustrations from a single distortion-free viewing angle, giving the book a cleaner, more organized, contemporary appearance with a comfortable flow from one illustration to the next.

Heights and measures presented another issue that needed to be treated with a little more consistency than it usually received. It is not unusual, for instance, for an illustrator to be given reference material showing mountains 50 miles high and oceans 20 miles deep. Working with the author, I determined the normal height and dimensional range for each of the formations and strata that would repeatedly appear in the book. This enabled me to keep important geological features in proper relational scale to each other.

A carpenter's projection, used throughout many geology texts, is a distortion of perspective.

CARPENTER'S PROJECTION

IN ORDER TO SEE THIS, THE TOP VIEW WOULD HAVE TO LOOK LIKE THIS

PROPERLY DRAWN PERSPECTIVE VIEW

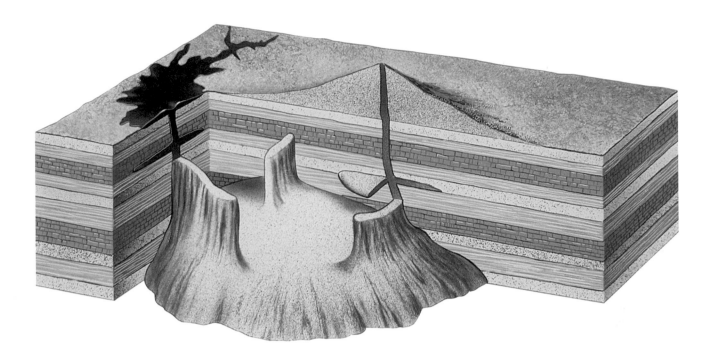

A rising molten mass, in red, is shown three-dimensionally in this cutaway view.

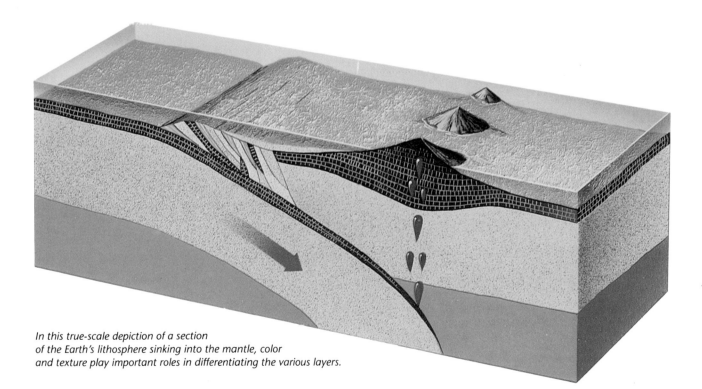

*In this true-scale depiction of a section
of the Earth's lithosphere sinking into the mantle, color
and texture play important roles in differentiating the various layers.*

COMBINING AIRBRUSH AND PAINTING TECHNIQUES

Geological subjects also lend themselves to combining approaches. Airbrush and hand-painting techniques, for example, can work very well in tandem.

Whereas aerial views of mountainous terrain can be done with side spraying and airbrush textures, the close-up views required for geological illustration necessitate that individual features be painted in by hand. The airbrush can still be used to show the cutaway sections of the Earth's strata, however, with differing airbrush textures defining the layering patterns. It makes no difference whether the hand painting is done before or after the airbrush work; personally, I prefer to do the exacting hand painting first and then mask it for spraying.

Here two layers of silicone putty—one dense and opaque, the other less dense and transparent— were spun in a centrifuge to simulate salt upwelling as it occurs naturally. The author supplied a top-view photo plus additional verbal reference information.

The haze just above the surface of the ice in this illustration of crevice formation in a glacial flow was achieved by airbrushing a thin mixture of zinc white over the canyon walls.

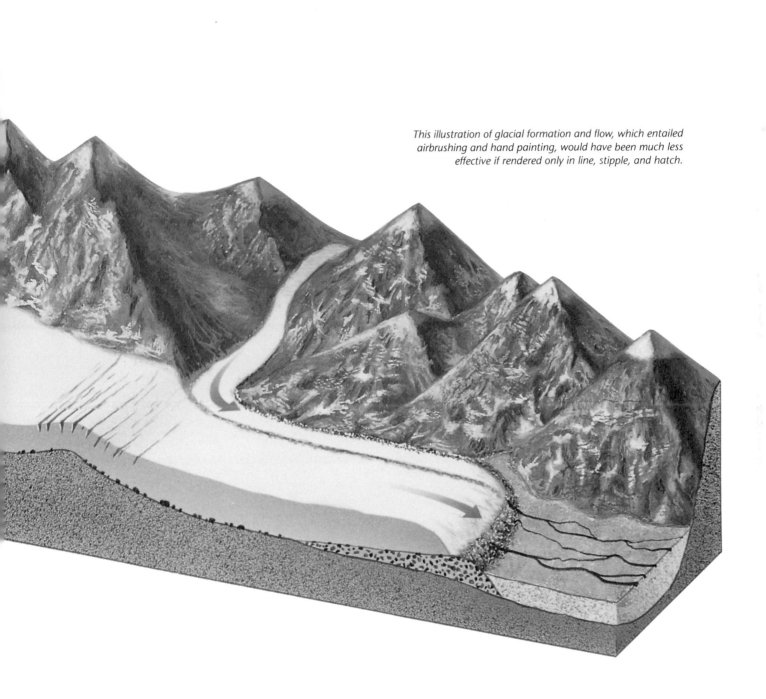

This illustration of glacial formation and flow, which entailed airbrushing and hand painting, would have been much less effective if rendered only in line, stipple, and hatch.

ARTICULATING THREE-DIMENSIONAL STRUCTURE AND CONTOUR

These three-dimensional drawings from an article on salt tectonics were done from computer-generated linear simulations by adding lighting, shading, and contouring. The computer-generated simulations work in much the same way as a topographical map—each computer line represents an elevation and a contour. The forms then have to be filled in and rendered to give the subjects solidity.

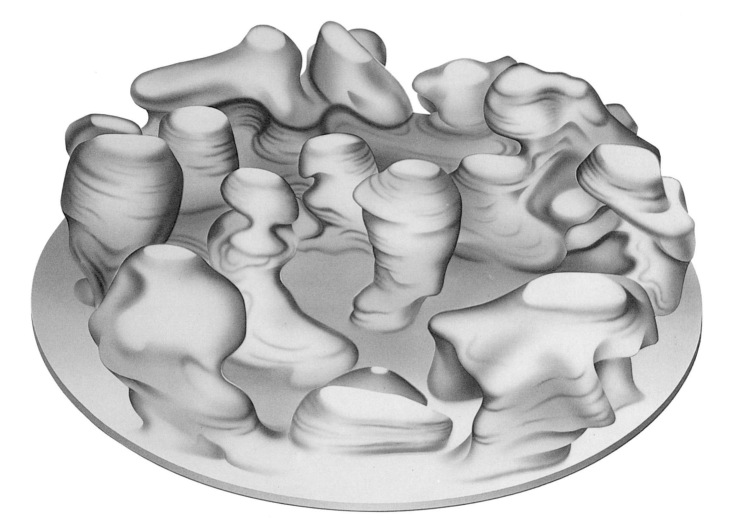

A computer-generated contour line drawing was the basic reference for this rendering of the distribution and shape of typical salt domes. Light is played against dark, allowing structural detail to be visible in every area. The effect is similar to that achieved by painting with light (see Chapter 4).

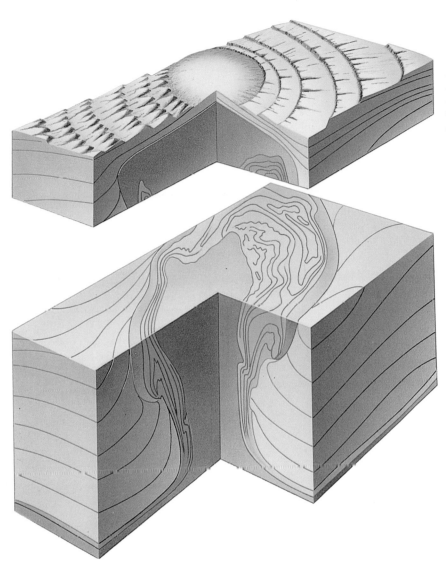

For this exploded view of a salt dome section, the airbrush work was followed by a spray of a diluted gum arabic solution to stabilize the surface of the illustration prior to the line work. I used gouache in a No. 6 wrico pen to draw in the lines.

Contrasting colors are used to show layering in this illustration of a sediment-distribution experiment. Note the mushroom stem, which provides a good example of hard-edge/soft-edge transition, helping to create the three-dimensional effect. The more typical approach to this sort of image—black line and tints—would not allow for this result.

CHAPTER NINE

ARCHITECTURAL AND STRUCTURAL RENDERING

Although architectural and structural renderings are not generally thought of in the context of scientific illustration, they nonetheless fall into the scientific illustrator's domain. In fact, it is the *structure* of a given scientific subject that is the focus of most scientific illustrations. So although shapes and forms may differ, the techniques associated with illustrating structural features, whether natural or man-made, remain fundamentally the same.

My approach to rendering these subjects is essentially as follows: I familiarize myself with the subject by reading reference material, analyzing blueprints, and, if possible, making on-site visits; I make rough sketches to crystallize my approach; if warranted, I build and photograph a model, and then I make the preliminary sketches go on to the final drawing.

Naturally, each project comes with its unique problems. Solving them is what makes the work interesting. Large, complex structures often present a problem in determining the perspective projection to be used for an illustration. In most cases, this can be done by accurately plotting the floor plan onto a perspective grid or by photographing the floor plans, as described in Chapter 4.

Cross-sectional view of a house designed to utilize solar energy as a heating source.

Bird's-Eye Views for Architectural Renderings: The White House

Rendering the working areas of the White House for *Fortune* magazine in 1975 presented a unique set of problems well outside the normal range of challenges. The magazine was preparing a feature on the White House's West Wing rooms, where the day-to-day business of government takes place. These areas are not open to the public, and this is the only illustration to define them for public view.

I prepared for an on-site visit and left for Washington as soon as my security clearance was arranged. A number of obstacles were immediately apparent, particularly time—the completed project was needed in a week. Worse, the only available floor plan was seriously inaccurate.

On the plane to Washington, I sketched possible viewing angles that would present an unobstructed view of the two floors of rooms, their furnishings, and connecting passageways. By the time the plane landed, I had settled on a bird's-eye view, which shows only two walls of any given room and therefore requires fewer reference photos.

My arrival coincided with a sudden international crisis, resulting in a media and security frenzy that made it difficult for me to gain access to rooms in the sequence I desired. To visualize the overall layout, I had to rely on the inaccurate floor plan and modify it as I went.

Entering each room, I oriented myself with Pennsylvania Avenue to establish a consistent reference point for the illustration. I photographed the appropriate walls with a Polaroid SX-70 camera, and used the camera's rangefinder to measure the length and width of each office. Beyond that, there was barely enough time for some close-up photos of the furnishings and a few quick sketches—five hours after my arrival, I was on my way back to the airport.

In my studio the next morning, I transferred the rangefinder measurements onto a floor plan drawn on illustration board, making corrections and adjustments as needed. Then I glued thin cardboard wall sections in place to complete a simple model.

Keeping Pennsylvania Avenue as the reference point, I tried different camera elevations and angles until I had a good view of every room without having to cut away too much wall area. In order to achieve a perspective that accurately showed the relative positions of the first and second floors, I propped the second floor in position above the first with balsa wood sticks, essentially creating an exploded view. By midday I had the reference photo I needed for a bird's-eye perspective drawing. I drew in the furniture, wall treatments, and other room details from the on-site Polaroid shots. The next day I delivered a detailed pencil sketch to *Fortune* for approval.

The final illustration was done almost entirely by airbrush; furniture, drapery, and detailing were painted by hand.

Reference photos taken in the White House. Clockwise from top left: the Chief Congressional Counselor's office; two shots of the private office of the President; a partial view of the Oval Office; lobby reception area; the Secretary of State's office.

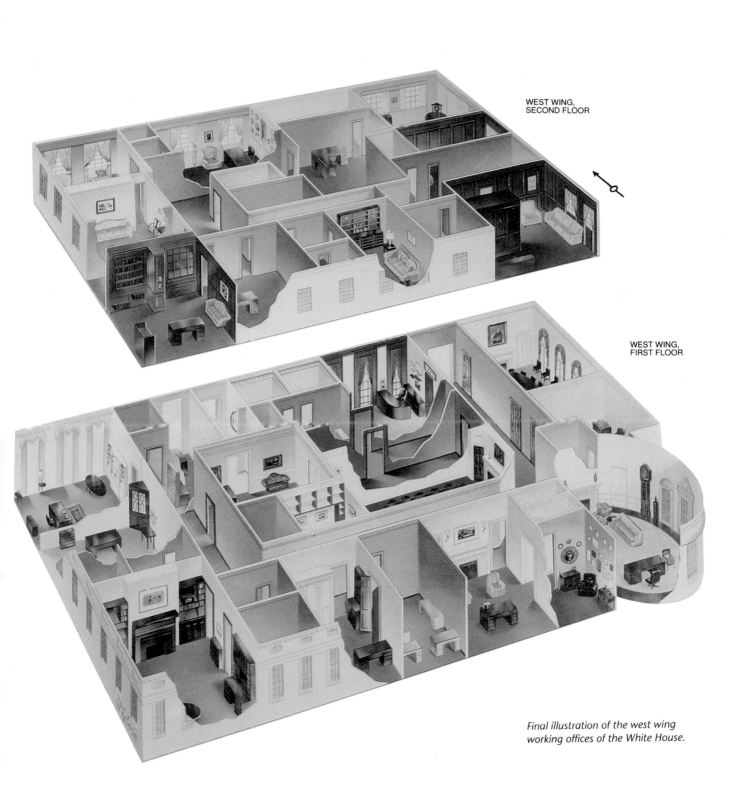

WEST WING,
SECOND FLOOR

WEST WING,
FIRST FLOOR

Final illustration of the west wing working offices of the White House.

DESIGNING AN APPEALING PROSPECTUS: THE PORT OF NEW YORK AUTHORITY BUILDING

With the completion of the World Trade Center in the early 1970s, the Port of New York Authority packed up its headquarters and moved into the new skyscrapers, leaving behind a vacant 15-story building that then had to be sold. Situated on an entire square city block, the massive, cavernous structure offered such a variety of technical and industrial facilities that its architectural blueprints proved to be too complicated for real estate brokers and potential buyers to deal with—in addition to indoor parking facilities and 600,000 square feet of office space, the building has street-level truck berths, a helicopter landing pad, 12 high-speed passenger elevators, 11 10,000-pound freight elevators, and four trailer truck–sized drive-in elevators for up to a dozen trucks per floor. It was decided that a full-color sales brochure would be an attractive and more practical means of presenting the building, and I was called in to create an illustration that would make it easier for prospective purchasers to visualize the extraordinary potential of the building's internal space.

A tour of the building convinced me that I would have to use an array of creative techniques if one illustration was to fulfill so many objectives. I needed to show a minimally sufficient level of detail while still projecting the total structure; this entailed as much time figuring out what to omit as what to include. Working with blueprints, floor plans, and the notes and photographs taken during my visit, I planned out my approach.

To show the street-level loading docks, the second through fifteenth floors were exploded upward. Cutting away a section of the first floor exposed the basement cargo loading facility. The building and elevator shafts were ghosted, providing an unobstructed view of a typical office floor. Color coding was used to define the individual working areas, and earthy colors were used against a golden glowing background to give the drab, industrial building an inviting sense of friendly warmth.

As it turns out, the building was sold soon after the publication of (and, I'd like to believe, with the help of) the brochure.

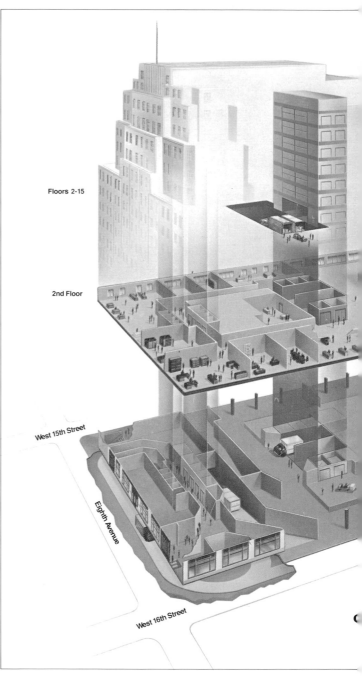

Floors 2-15

2nd Floor

West 15th Street

Eighth Avenue

West 16th Street

Aerial photograph of the building.

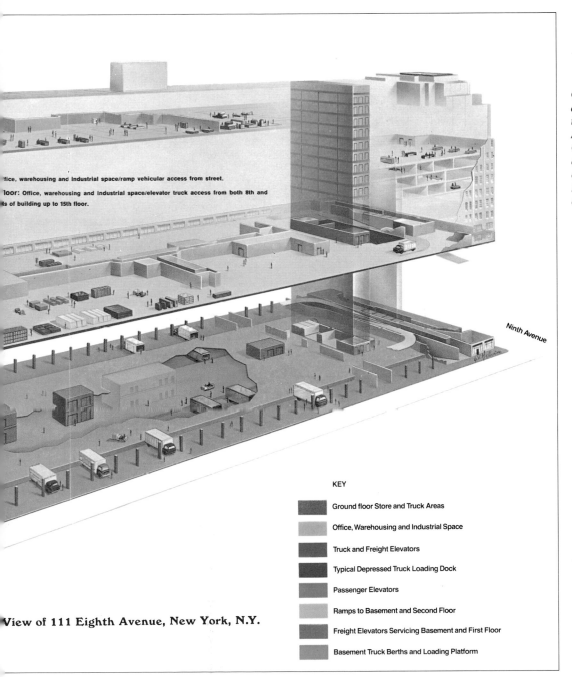

*fice, warehousing and industrial space/ramp vehicular access from street.

loor: Office, warehousing and industrial space/elevator truck access from both 8th and
s of building up to 15th floor.

*Combination cutaway/
exploded/phantom view of
the old Port of New York
Authority building.
Columns and other
architectural details that
obscured more important
building features were
omitted.*

Ninth Avenue

KEY

Ground floor Store and Truck Areas

Office, Warehousing and Industrial Space

Truck and Freight Elevators

Typical Depressed Truck Loading Dock

Passenger Elevators

Ramps to Basement and Second Floor

Freight Elevators Servicing Basement and First Floor

Basement Truck Berths and Loading Platform

View of 111 Eighth Avenue, New York, N.Y.

DELINEATING COMPLEX SPACE: THE WORLD TRADE CENTER

Soaring 1,350 feet (411 meters) above a 5-acre (2-hectare), parklike plaza, the two skyscrapers of the World Trade Center stand as a truly remarkable engineering feat. Shortly before construction of the Twin Towers began, I was asked to prepare a series of illustrations outlining the special features and facilities planned for the great concourse level. The Port of New York Authority, which wanted to use these illustrations for showing the rentable office and commercial space that would become available, specified that the illustrations had to describe the towers' relationships to commercial spaces, entrances, walkways, escalators, elevators, trains, and streets. In essence it was to be a "people flowchart."

Since most people find architectural and engineering drawings difficult to follow, I chose to utilize a bird's-eye perspective view instead. First, blueprints showing the concourse level's entire floor plan had to be photographed in perspec-

tive (see Chapter 4). The photos then became the basis of a comprehensive pencil layout. Once it was approved, I prepared a base illustration of the entire promenade complex.

I proceeded to delineate the various space configurations of this huge area through a series of overlays. Ten percent gray acetate overlay sheets were used to tone down surrounding sections, and windows were cut into the acetate in order to highlight specific areas. In all, 20 different overlays were required. Using the one piece of base art in conjunction with each of the gray overlays, I prepared 20 color slides detailing the available rental space.

A different procedure was employed for drawing the World Trade Center's multileveled Commodity Exchanges Building. The trading floor, rental space, and support areas were shown by combining a cutaway and exploded view, color-keyed to the various sections.

The base illustration of the promenade depicted a large amount of complex space, making it difficult for viewers to focus on any single area. This problem was solved through the use of windows cut in overlays of 10-percent gray acetate, which served to highlight specific areas.

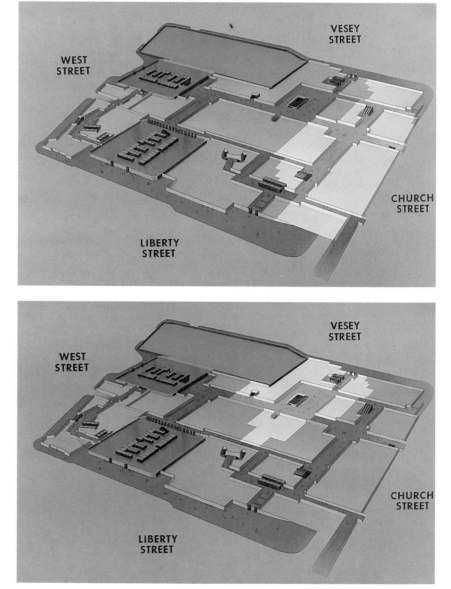

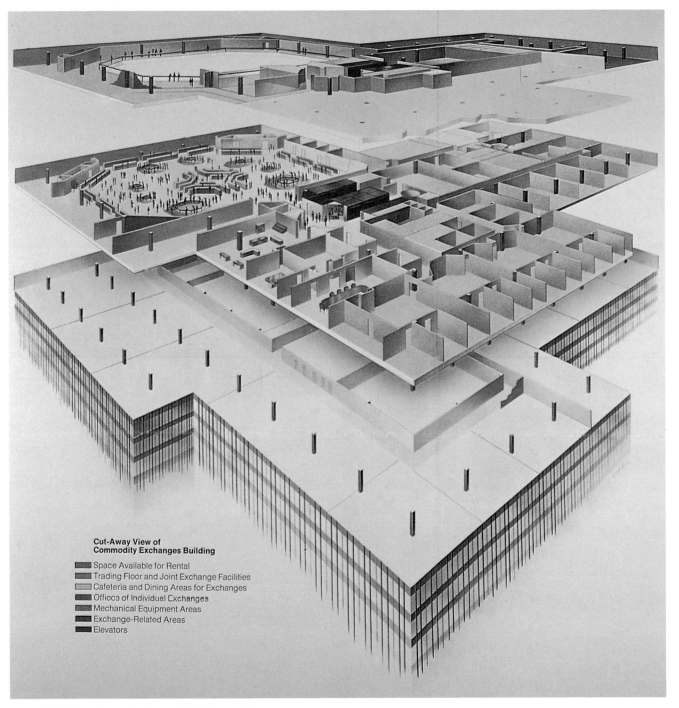

**Cut-Away View of
Commodity Exchanges Building**

- Space Available for Rental
- Trading Floor and Joint Exchange Facilities
- Cafeteria and Dining Areas for Exchanges
- Offices of Individual Exchanges
- Mechanical Equipment Areas
- Exchange-Related Areas
- Elevators

*An exploded view shows the Commodity Exchange's office space
and trading floor, all color-coded for easy identification.*

The second part of this assignment took me in an entirely different direction: Due to the unprecedented interest the construction of the World Trade Center was generating, a fully illustrated 16-page booklet was prepared and given away to the ever-increasing number of "sidewalk engineers" who gathered daily to watch the construction process. Each page featured a detailed drawing scaled from blueprints, with supportive text explaining the main aspects of the construction, starting with the initial excavation.

This aerial view of one of the towers shows construction taking place atop the structure as well as on the ground. A loading pier and a 23-acre (9.3-hectare) landfill, built with material from the excavation, can also be seen.

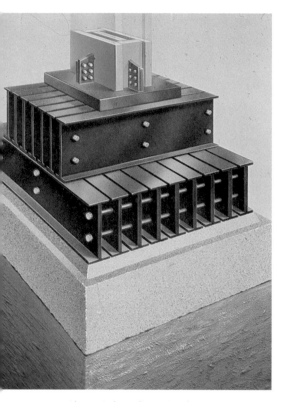

Above: A three-dimensional view gives emphasis to the massive steel grillage of the World Trade Center's concrete foundation.

A combination of airbrush and painting techniques were used to create a realistic painting of a prefabricated floor section being lowered into place.

Load-bearing columns are hoisted into position. No background is shown, but workers serve to indicate scale.

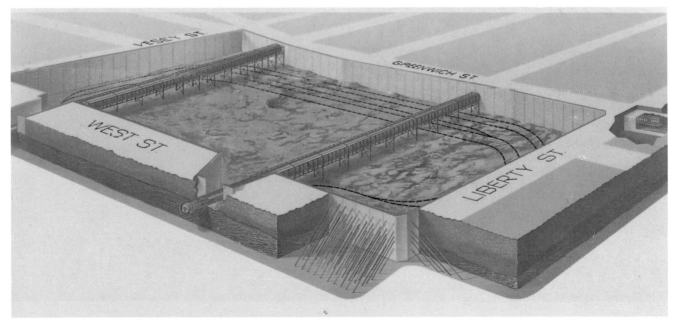

This bird's-eye view shows the tower's entire excavation site. The corner of the foundation wall was cut away to expose the supporting tie rods and the subway. Dotted lines indicate tracks for the new New York–New Jersey underground train link.

Giving Structure to an Environmental Concept: Energy Island

Nigel Chattey is an engineer and a man with a vision. His dream is to build an "energy island" in the Atlantic Ocean, 20 miles off the coast of New York and New Jersey—a multipurpose island providing a world-class deep-water port as well as a wide range of heavy-industrial, waste-management, and energy plants. In concept, it would solve many of the environmental problems facing us today.

I met Chattey in the early 1980s while on assignment for *Next*, a scientific features magazine, which was preparing an article describing the principle and extraordinary benefits of such an island. The project had already been under discussion for several years, and there was now hope to develop public support and bring it to fruition. It was thought that a comprehensive illustration would help the effort.

I worked primarily from photographs of the offshore deep-water port built by the Dutch off the coast of Rotterdam, the Netherlands, which is conceptually similar to Chattey's brainchild. The rest of the reference information came from working closely with Chattey himself, who carefully furnished me with the specific dimensions and descriptive details of his plan.

I used a perspective chart to maintain exact scale, each square representing 100 feet (30.5 meters). A cutaway of the ocean provides a cross-sectional view of the sea wall. Turbulent waters shown outside the sea wall are in contrast to the calm channels on the inside, which were designed to ease docking and unloading of tankers and ships. Although these and the other modular components appear to be relatively small in the final illustration, they are shown in correct scale to the island.

Just a glimpse of Manhattan's skyline is visible on the horizon. A light, transparent spray of zinc white gives the sky its characteristic hazy look. This technique creates an effect similar to the glazing method the old masters employed to give their paintings an atmospheric quality. Original art: 16 × 20" (41 × 51 cm).

Re-Creating an Ancient Warship: The Trireme

In contrast to the Energy Island project, in which I had to depict something that did not yet exist, the assignment to render this ancient Greek warship—a trireme—essentially called for me to re-create something that no longer existed. The vessel was powered by 170 closely packed oarsmen sitting on three levels of benches. The authors of a *Scientific American* article on boat's design and construction prepared a collection of rough drawings and calculations, as well as information on the ship's dimensions and the seating arrangement of the oarsmen.

From this limited information, I had the seemingly reasonable impression that all the oars were the same length, but something about that premise bothered me. In my mind I pictured the top row of oars projecting out from the side of the ship—since they were the farthest from the water, wouldn't they strike the lower oars if they were all the same length? To find out, I punched some holes corresponding to the oar holes into a piece of cardboard. By placing equal-lengthed balsa wood dowels into the holes and taping them together, I was able to "row the oars" in unison. Given the spacing and stacking arrangement of the oarsmen, this proved that the oars would have to be different lengths in order to clear each other. The authors were happy to have this piece of the puzzle solved, and I proceeded to my pencil sketch.

Using a perspective grid, I converted the authors' dimensions and structural information into a line drawing, cutting away the midsection to reveal the inner construction and outboard rigging. Due to the drawing's complex detailing, I added a color tint on the inside of the hull to clearly differentiate it from the outside surface.

A stone relief carving showing a trireme, c. late fifth century B.C., kept at the Acropolis Museum in Athens, Greece.

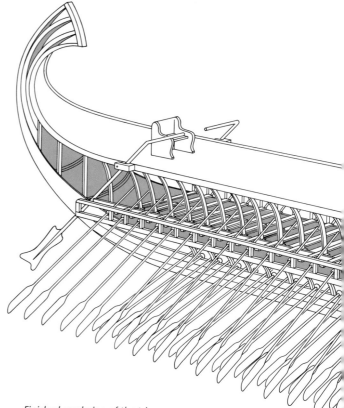

Finished rendering of the trireme.

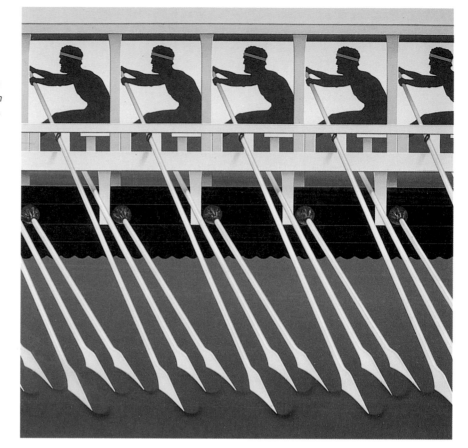

This decorative, stylized painting was created for the magazine cover, using simple shapes and a repetitive pattern to focus interest on the overall design.

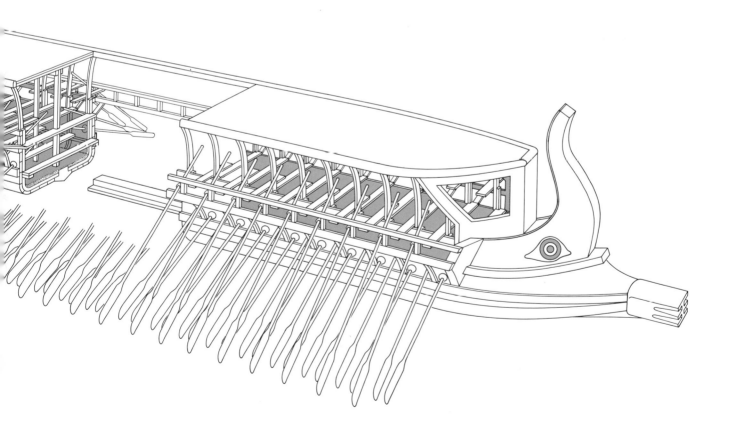

Expressing Tremendous Size and Volume: An Oil Supertanker

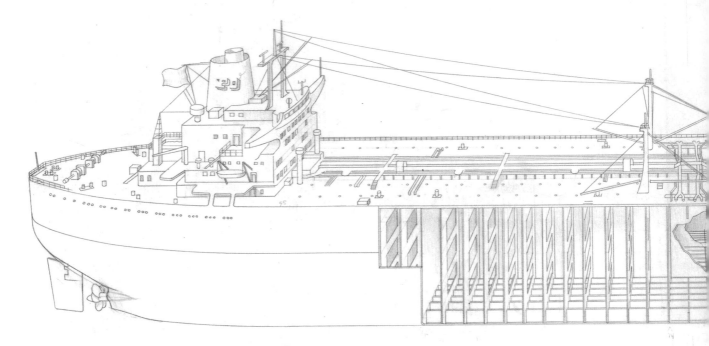

From old to new: The trireme project, involving an ancient seagoing ship, forms an interesting juxtaposition with my mid-1970s assignment to render a modern oil supertanker. The painting was commissioned by Readers Digest Books for a nonfiction story about supertankers in world commerce, and this time, thankfully, blueprint drawings and photos were readily available.

The overall view focuses attention on the tanker's tremendous size and huge storage capabilities. The center section is shown empty, revealing the inner structure of this mammoth vessel, measuring 900 feet (275 meters) long and almost 100 feet (30 meters) wide. Backlighting the center section provides an effective dramatic touch. In a separate close-up view, a cutaway exposes the bridge, engine room, and crew's quarters.

All the featured structures of the tanker were converted to the correct scale and plotted onto a perspective grid. The base drawing was accomplished by correlating references from three different sources. After making a drawing of one bulkhead from blueprints (see cross section, shown in inset), the dimensional information was plotted onto the perspective grid and duplicated for the remaining bulkheads. The rest of the drawing was plotted in the same fashion. Photographs were used as reference for the bridge and engine room and a small orthographic drawing serving as reference for the deck area.

Above: Pencil drawing of a supertanker in one-point perspective. All lines on horizontal planes converge to a single point at the horizon for a three-dimensional view.

Top right: Full-color rendering of the pencil drawing. Left and right bulkheads are half-filled with oil; center section is shown empty to indicate full depth capacity of the storage area.

Bottom right: Painting of the aft section. Cutaway of the bridge command center shows the operational facility; the engine room and maintenance area are shown below. The small inset illustration details the configuration of the steel bulkheads in the oil storage section.

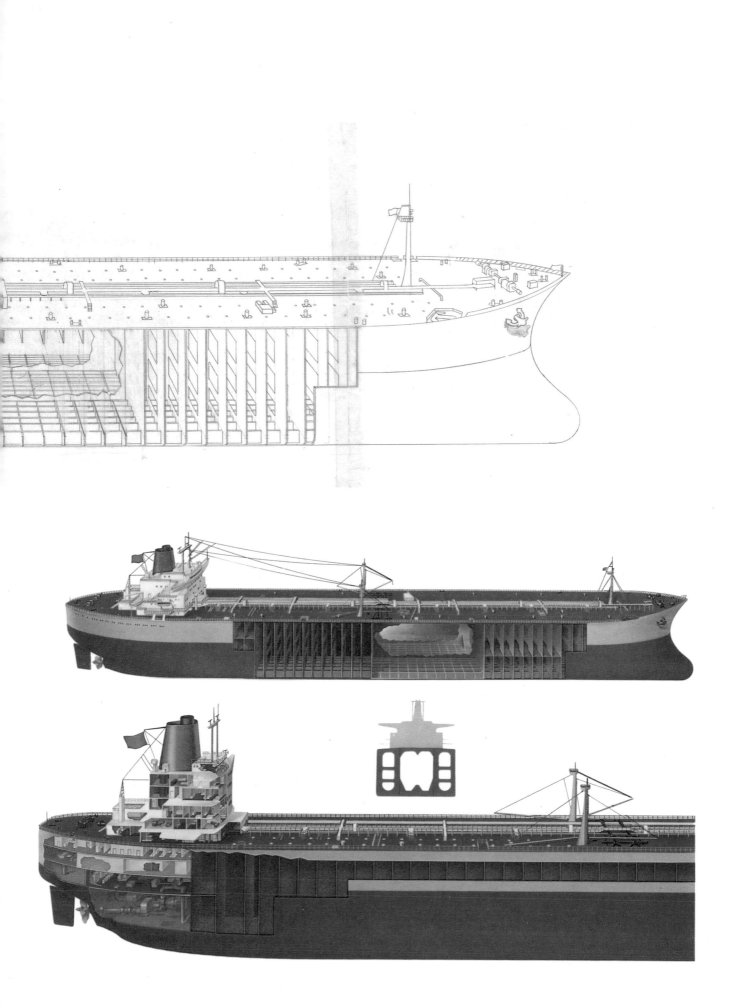

COLOR-CODING IN A DETAILED DESIGN: THE FRANCE

For a Time Inc. book, *Ships*, I undertook the preparation of a detailed cutaway view of an ocean liner to introduce an essay extolling the advanced engineering and unparalleled elegance of these floating cities.

The ideal subject was the *France*, the most modern and luxurious ship of the 1960s. The editor's initial plan was to use a painting of the ship that had been obtained from the cruise line. However, it was a somewhat stylized version, with deep shadows and colors close in value and chroma. When it was reduced to book size, the features became indistinguishable.

My assignment was to create a new painting, with the first one as my reference. Since the publisher intended the image to appear as an impressive double-foldout spread measuring 32 inches (81 centimeters) across, a substantial amount of interior detail was planned for inclusion. The completed painting, measuring 75 inches (1.9 meters) long—my largest single illustration—was probably the most difficult and intensive project I ever took on.

A great deal of work had to be done in order to adapt the original painting to this new format. Perspective had to be corrected, and most of the blurred images in the original had to be redrawn. While the assignment presented a challenge, and I was anxious to do it, the publisher's unrealistic 30-day deadline meant I would have to plan the project carefully, since the tight time restriction would not allow for corrections.

I began work on a pencil sketch, drawing in as many of the fine details as possible. Every feature of the ship that a cutaway could reveal was represented, including stateroom furnishings, cargo, and crew quarters—I left nothing out. The completed sketch provided a color-coding system for identifying and locating the various parts of the ship, a feature that was missing from the original painting.

The color palette was selected so that the colors would complement each other yet remain unmistakably distinct. Since this would be a shadowless illustration, I worked with three values of each color to define three-dimensional structures; dark colors were used sparingly, and black not at all. Eight separate color schemes were employed, each used to indicate a different category of space—greens to denote staterooms, golds and browns for engine and boiler rooms, and so on. I selected a more realistic, natural-looking color palette for the public rooms.

By the time the preliminary drawing was approved, only 21 days remained to complete the painting. I divided the drawing into 21 segments and planned to finish one each day. By working 10 to 12 hours per day, I managed to keep to the schedule and deliver the work on time. The painting hangs on the wall in my den now. Looking at it still gives me a lot of personal satisfaction, but one assignment like this in a lifetime is enough.

Cross-section airbrush rendering of the luxury ship France, *drawn to accurate scale and extensively detailed. Original art: 18 × 75" (45 × 190 cm).*

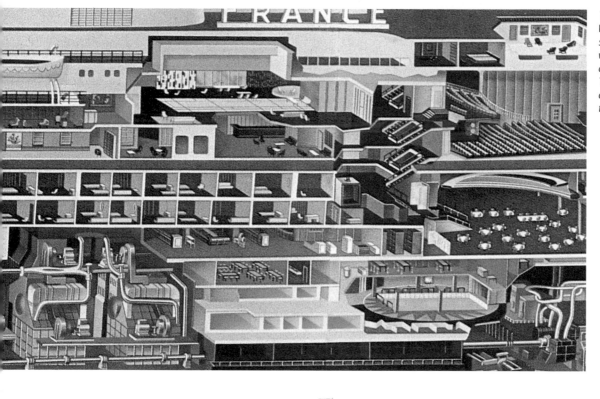

This detail of the France *illustration shows the size at which the image was actually reproduced. The publisher used a double-foldout spread to accommodate it.*

CHAPTER TEN

ASTRONOMY AND SPACE TECHNOLOGY

Our collective quest to discover the truth about ourselves, who we are, where we came from, and where we are heading has been an enduring preoccupation since the earliest times. Even today, we continue to search the skies for the answers to these questions. Thanks to the genius of Galileo, Einstein, and the many others who have contributed so much to this pursuit, we have reached far beyond the limits of our own galaxy and now find ourselves with the means to explore the outer reaches of the universe.

And more wonders beckon: Observations are no longer Earthbound. Today's spacecraft carry sophisticated optical and electromagnetic detectors capable of scanning for emissions from space and transmitting them back to Earth. Plans for the future include a permanent observatory on the moon, equipped with optical and radio telescopes to examine the entire electromagnetic energy spectrum of the universe. These topics will undoubtedly be the subjects of many scientific illustrations in the years to come.

Astronomical observations indicate that the universe has an uneven distribution of matter. The translucent spheres in this illustration define the limits of matter surrounding each galaxy.

VISUALIZING THE MOON'S MAGNETIC FIELD

I am sure you remember the grade school experiment of placing a bar magnet beneath a piece of paper and sprinkling iron filings on top. The filings immediately move toward the magnet's poles, thus revealing the pattern of its magnetic field.

Some scientists theorize that our moon may have had its own magnetic field billions of years ago, generated from a core of molten iron. I was assigned to illustrate this theoretical magnetic field for the cover of *Scientific American*.

The illustration has three elements: the moon, its core, and its magnetic field. I decided the core could be shown best by treating the moon as a semitranslucent sphere; representing the magnetic field, however, posed a larger challenge. Although the field would have been structurally similar to the one generated by a bar magnet, imitating the pattern of the iron filings in a three-dimensional representation would be unworkable due to the density and complexity of the pattern—the lunar surface and internal iron core would be obscured. Instead, I opted to show the field as lines emanating from the moon's magnetic poles.

Using the filings as a guide, I drew the outline of three generic magnetic field lines. Preliminary thumbnail sketches indicated that these lines could be represented effectively on three intersecting planes and that six field lines per plane would be optimum, conveying the story's essence without the confusion of extraneous field lines and detail.

Had I owned a computer at the time, I would have used it to plot the curves in perspective. Instead, I built a model from a 1.5-by-1.5-by-6-inch (3.8-by-3.8-by-15-centimeter) block of wood, into which I drilled a series of holes on three intersecting lines. Using the filings pattern as my guide, I cut three lengths of one-sixteenth-inch (16-millimeter) flexible plastic rod and approximated the curve of the magnetic field. These three rods then served as the pattern for 15 more, giving me a total of 18—six for each plane. The rods were then inserted into the wood block and supported with copper wire stiffeners.

To make sure the thin rods would photograph well, I sprayed the model black, placed it against a white background, and hung it from the ceiling by a piece of string before taking the photos. The cover design specified a 7-inch (18-centimeter) square for the final image, so I enlarged the best photo to work at twice up from that size.

In the first sketch I prepared, I illuminated the moon from above, which I felt would give the illustration a more dramatic effect. In so doing, however, I inadvertently introduced an error into the image and only the alert eye of the project editor saved the illustration from a serious misrepresentation. By calculating the relative position of the Earth to the moon, it was determined that the moon's magnetic poles were oriented similarly to those of the Earth, which meant the sun would be to the side of the moon, rather than above it. The corrected view, with the proper illumination, can be seen on page 161.

Nine masks were needed for the airbrush painting of the field lines. Each one was cut and used for the left side of the painting and then flipped over to draw its mirror image on the right. The masks were made by scoring the field lines into a sheet of acetate so that the acetate could be separated into two segments, allowing me to adjust the width of the field line.

I began by painting the illustration board black, and then painted the moon and its core over the black background. The field lines were another matter—with the masks in position, I airbrushed white over the black background followed by orange paint over the white, a procedure that gives the most brilliance. Had the base color been left black without the intervening coat of white, the color separator's scanner would have detected it and the orange would not have reproduced as well. This can be seen by comparing the color of the lunar core with the more reflective brilliance of the magnetic field lines.

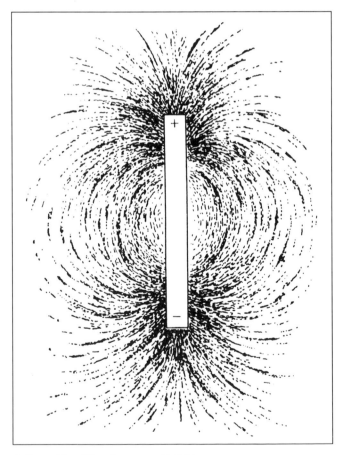

Pattern of iron filings in a magnetic field.

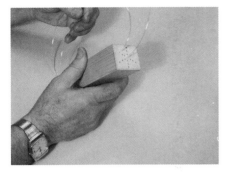

Thin plastic rods, inserted into predrilled holes in a block of wood, were used to represent the moon's magnetic field.

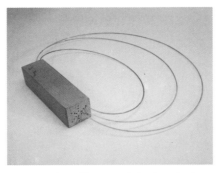

Three graduated lengths of plastic rods after insertion into the block of wood. Note the similarity to the filings pattern.

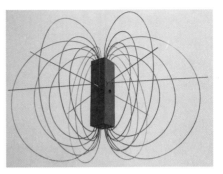

Completed model suspended by a string. This photograph was used as reference for the final illustration.

Final illustration of the lunar magnetic field.

COSMIC PHENOMENA

Much of the phenomena taking place in the cosmos can be detected but not observed. Gamma rays and black holes, for example, definitely exist, as various measurements clearly attest, but we cannot *see* them. Subjects such as these present a special challenge—how is the artist to know what these phenomena look like, and how to depict them?

Reference data for these projects, usually supplied by the authors, takes various forms; sketches, text, and computer-generated graphics are the most common. Together, the author and artist must transform this raw data into a recognizable visualization that accurately describes the phenomenon. For example, solar wind—the ionized atoms and electrons generated by the sun—might reasonably be depicted as looking something like grains of sand streaming away from the sun's surface. The particles farthest from the sun would have the longest trails of light; those nearest, the shortest. This sort of logical thinking allows the artist to re-create the action and physical attributes of the subject.

Airbrush techniques are especially adaptable to astronomical illustrations. Soft gossamer effects are the major characteristic of airbrush spray, and therefore can be achieved without masking. In order to convey some of the vastness of outer space, these images were painted directly against black backgrounds—a departure from the approach taken with most other types of subject matter, in which the images are masked out to retain the underlying reflective quality of the white board.

With no confining, finite borders to consider, all structural edges are gently faded out, and only schematic diagrammatic representations are depicted with hard edges. Continuous-tone freehand spraying effectively simulates gaseous phenomena, a fine stipple can be used to replicate granular textures (as in the spiral arms of galactic formations), and coarser stipple works well to suggest a background star field.

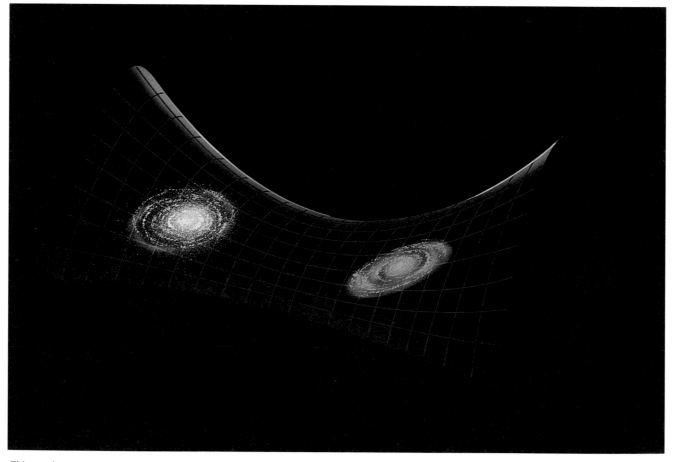

This purely conceptual painting of the geometry of an open universe, painted on a black background, uses only enough shading to bring out the saddle form. Galaxies are the prominent features; grid lines, which help to define the saddle's shape, were drawn with the aid of the brush stylus (see Chapter 6).

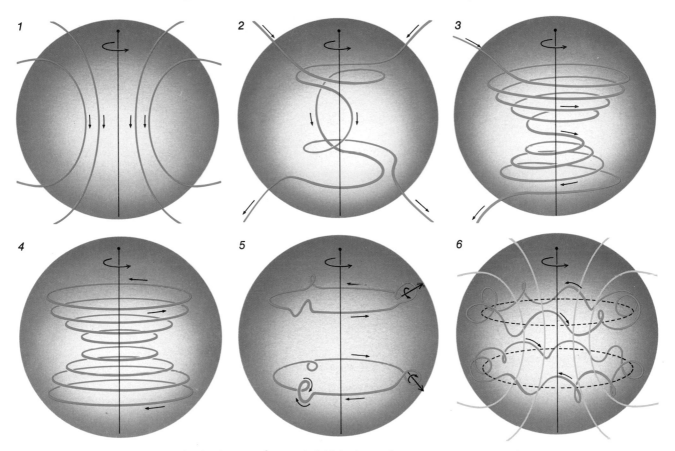

This step by step progression shows the development of magnetic fields in the sun. In contrast to the Earth's solid structure, the sun's gaseous composition and ever-moving currents trap magnetic field lines, causing the distortions shown here.

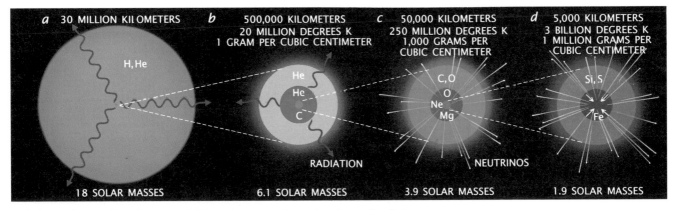

Color is used here to show the stages of this pre-supernova star undergoing nuclear fusion. The center portion of each diagram is enlarged at right to show greater detail and solve the problem of scale.

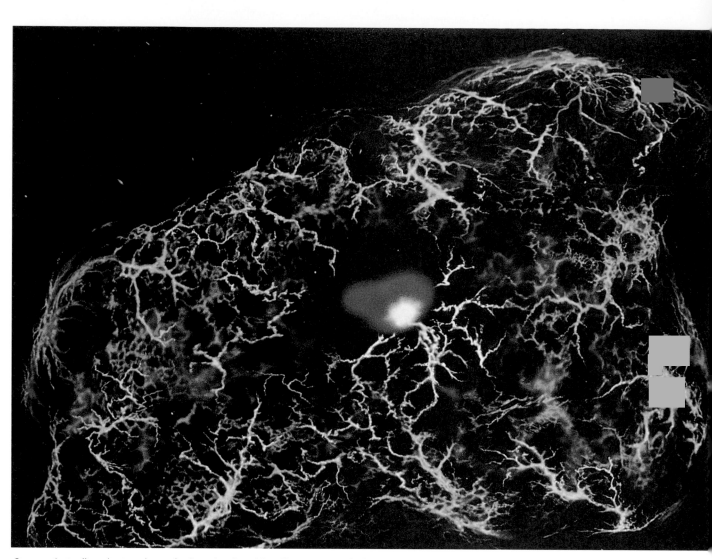

Quasars (actually a shortened term for "quasi-stellar objects") are among the largest structures and strongest energy radiators in the universe. But because they exist in the farthest reaches of the universe, they appear only as faint points of light when viewed through even the most powerful telescopes. For this theoretical "portrait" of a quasar, I worked from reference photos of galaxies and star formations, supplemented by the author's verbal description of the filamentous outer structure. My objective was to convey the feeling that we were looking at it through a giant globular shell. To do this, I muted the background filaments in both color and value and progressively lightened the values as I worked toward the foreground until I was finally working with pure white.

Particles shooting out from the sun—
a phenomenon known as the solar
wind—shape the Van Allen radiation
belts surrounding the Earth. The brush
stylus (see Chapter 6) and airbrush
spattering were used to achieve the
effect of the streaking particles. A
cutaway view shows a profile of each
shell-like radiation belt.

This freehand airbrush painting shows
the symbiotic gas exchange between
two stars and their extended,
interacting arms. The elliptical
background of stars was airbrush-
spattered through an elliptical
cardboard mask, which was held above
the surface to achieve a soft edge.

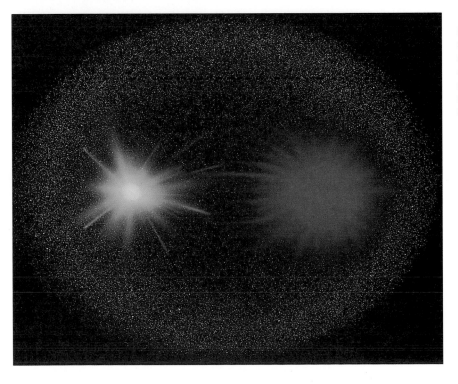

CAPTURING THE LIFE CYCLE OF COMETS AND METEORS

Comets and meteors, although often mistakenly referred to interchangeably, are quite different from one another. A comet seen with the naked eye typically appears as a faint glow of light, often with a diaphanous tail streaming away from its head. Seemingly fixed in the sky like the moon, stars, and planets, it slowly changes its visible position over a period of a few days to a few months. The tail, which consists of gases and dust, always points away from the sun—not back along its orbital path, as one might expect.

As the comet's nucleus evaporates, its dust particles scatter about the solar system. The larger particles reaching the Earth's atmosphere are meteors, destined to burn up from the heat of atmospheric friction, giving off the brief, visible flash we see as "shooting stars" in the night sky.

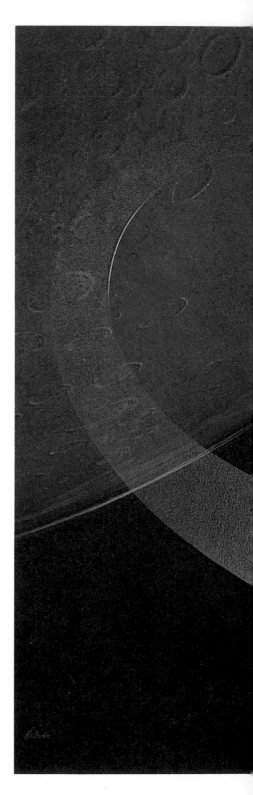

In the dramatic painting above, emphasis is directed to the path of a speeding comet as it emerges above the orbital plane of our solar system. The near sides of the orbital ellipses are thicker and brighter, creating a luminous quality and suggesting the feeling of great distance between the near and far sides of the orbit.

The illustration at right describes a theoretical sequence of events in the life of a meteor: An object striking Mars causes surface material to fly into space, where it collides with another body, causing particles to explode out in all directions. These particles are now technically meteors, some of which become trapped by the Earth's gravity and eventually burn up in our atmosphere.

The curving, sweeping pathways convey the sequence of events in a dynamic graphic design. Playing the curves against each other gives the eye an interesting path to follow while creating an illusion of space. I used color to distinguish between the events and left the first section of the pathway white to accommodate the title of the article.

ROCKETS AND MISSILES

Rockets and missiles are recurring subjects for scientific illustration, so finding new and innovative ways to present them is a challenge. Rocketry can be traced back as far as ancient Asia, where early rockets, made by packing saltpeter and charcoal into bamboo shoots, were used as fireworks. As rocket technology has moved forward, so has public fascination with it. During the early years of the space program, rocket illustration was limited primarily to technical manuals and science magazines; with today's more sophisticated rockets serving as the backbone of our high-profile modern space program and as the vehicles for our nuclear arsenal—two topics of widespread public interest and concern—newspapers and general-interest magazines use rocket images with much greater frequency.

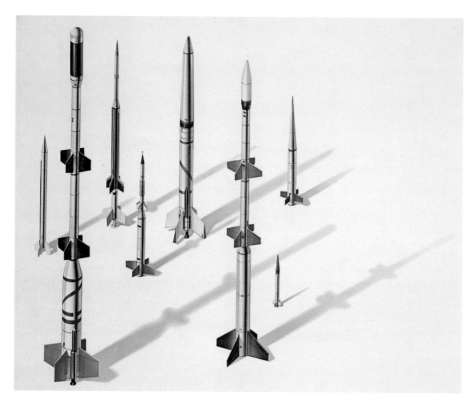

These early-1950s sounding rockets, ranging from 8 to 62 feet (2.4 to 18.9 meters) in height, were the original explorers of the Earth's upper atmosphere.

For this display of American and Russian rockets and missiles, various reference photos and drawings were scaled to allow the subjects to appear in correct proportion to one another. The brush stylus was used extensively, and the light-to-soft transition technique was employed to create the exhaust flames (see pages 88 and 94, respectively).

All of the projectiles are converging on a single point, creating a powerful composition. Titles and text were later surprinted in the light area of the sky, creating a light–dark dynamic that further heightened the effect.

COLOR-ENHANCED IMAGERY

This space vehicle, developed for the unmanned exploration of Mars, features interchangeable accessory packages designed for selective use in accordance with the goals of the mission. The initial plan was to show all the design variations as separate illustrations, but I proposed showing all the components in one illustration, with the alternate equipment packages floating in space near to where they coupled to the main vehicle. This plan, I felt, offered greater visual impact and made the concept easier to follow.

To increase the imagery's effect, the red glow of Mars is exaggerated in the reflective surfaces of the craft. In reality, most objects in space appear in high-contrast black and white, with almost no evidence of color, and are realistically illustrated with cool whites, blues, and shades of black. Blues and reds are used here strictly as a graphic device: The surfaces facing Mars appear to be reflecting its red color; surfaces that should actually appear almost black were rendered in shades of light grayish blue, in keeping with the colors commonly associated with space.

The soft edges of Mars were airbrushed through a cutout section of an acetate mask held above the painting's surface. Note the difference in brilliance between the red of the planet, which was rendered over the black background, and the red in the spacecraft, which was rendered over white.

The Mariner *spacecraft shown illuminated by the sun. Creative enhancement of color adds to the luminous effect.*

ELIMINATING THE NONESSENTIAL

If I were to illustrate the elevator system in the Eiffel Tower, I would have to leave out much of the structure's lacelike framework in order to reveal the core that houses the elevator itself. The same principle applies to all complex structures, whether they are buildings, telescopes, engines, or spaceships—illustrating intricate objects such as these requires a judgment call regarding which elements are nonessential and can therefore be sacrificed. Sometimes this entails not showing the full extent of these lesser elements; in other cases, it means eliminating them entirely. The basic principle, however, is the same: Meeting the objectives of an illustration generally means knowing what to leave out.

In this illustration of the Mount Palomar telescope in California, much of the supporting framework has been eliminated and the housing in the background has been blacked out in order to concentrate attention on the path taken by light passing through the telescope.

In this illustration of a multimirror telescope at Mount Hopkins, Arizona, it was necessary to show the path of light in only one mirror in order to convey the information of all six. This approach made it possible to communicate the illustration's basic scientific point—light reflection—without obscuring any of the structure's detail.

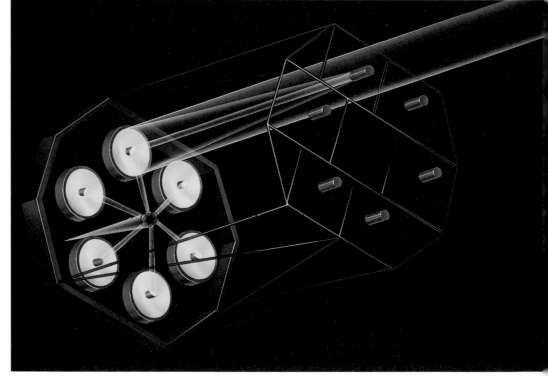

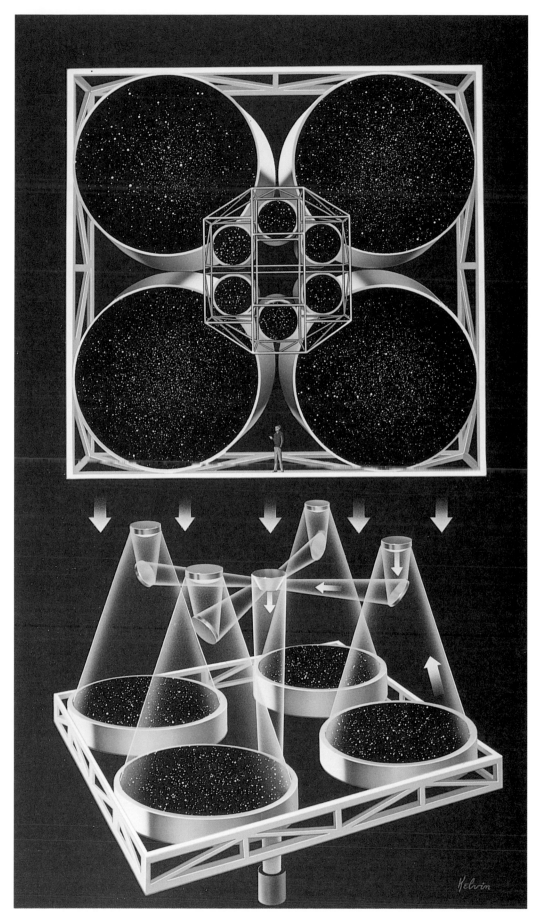

Here an existing telescope (shown in green) is juxtaposed above a proposed giant, four-mirror telescope to show the size differential between them. The perspective view below concentrates on the telescope's light-gathering paths. In both views, the support structures have been greatly simplified. The network of trusses and metal beams supporting the smaller, secondary set of mirrors, for example, was completely eliminated, so as not to interfere with the cone-shaped light pathways. The square mounting surrounding the primary mirrors was included, however, to indicate the relationship of the mirrors. Star patterns in the mirrors were simulated with airbrush spattering.

CHAPTER ELEVEN

SCIENTIFIC AMERICAN

MARCH 1990
$2.95

TRENDS IN COMMUNICATIONS: *The Road to the Global Village.*

Earthquakes: should the Midwest brace for a big one?

Physicists predict—and find—new radioactivities.

Observatories on the moon: *a bold proposal
for a practical lunar scientific outpost.*

This cover depicting an observation
base on the moon is another
example of the results of extensive
consultation and research that led
to excellent reference material. I
merely needed to position the
sophisticated receiving equipment
appropriately on the lunar
landscape—the reader's
imagination could be counted upon
to do the rest. (For images from the
article itself, see Chapter 5.)

SCIENTIF AMERIC

Parallel programming: seven-league b

Spare that rabbit—cell tests for gaugi

Cracking Maya writing: new window o

**Supernova life
from a s

For this cover, my job was to use a comprehensive
package of research data to reconstruct the life
cycle of a supernova, which spans millions of years.
I designed the illustration as a free-flowing
S-shaped curve to impart some sense of the time
sequence involved. Note the color progression as
the supernova ages: Although reds, oranges, and
yellows are generally thought of as hot colors,
while blues, greens, and violets are commonly
thought of as cool, the opposite applies to
astronomical phenomena. Blues indicate the hotter
stages of the star's life; reds, the cooler ones.

DESIGNING COMPELLING COVER ILLUSTRATIONS

Cover design plays a critical role in the success or failure of every magazine—nowhere is the development of a design concept of greater consequence than when it must pass the scrutiny of the scientifically oriented reader. What makes a cover illustration work? It must have immediate visual impact; it must tell its story quickly; it must be attractive. Every cover is designed with these essential criteria in mind, but it is the artist's individual approach to a project that gives a cover the personality and identity that ultimately makes it work (or not work, as the case might be).

In designing a cover illustration to fit a given publication's format and focus, I try to consider the client's personal taste and direction, as well as the intended market. I use pencil sketches to formulate possible approaches, sometimes aiming for the bold and dramatic, at other times setting my sights on the dignified and laid-back, depending on the general tone of the publication.

In addition to these considerations, the format of the publication often presents problems that are separate from those pertaining to the art itself. The style of such elements as the publication's logo, the type that has to appear with the art, the size, style and color of titles and blurbs, and so forth, is often kept constant from issue to issue in order to profit from the recognition value they accrue over time.

Every illustration requires care and research, of course, but this caveat is particularly applicable to cover illustrations. The two *Scientific American* cover images shown on this spread, for instance, entailed hours of consultation with the art directors and editorial staff. Ultimately, this type of preliminary work makes the illustrator's job easier, since the up-front preparation usually means that the final illustration itself can be conceived and planned with few problems.

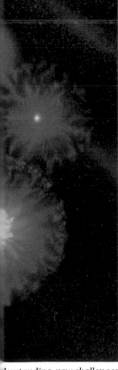

AUGUST 1989
$2.95

e mind.
sks.
nt culture.

derstanding, new challenges
n our cosmic neighborhood.

SINGLE-FOCUS AND SINGLE-TOPIC COVERS

A *single-focus* cover develops one element of a subject as the focal point of the illustration. It usually calls for bold use of color and exaggeration of shapes and proportions in order to create an interesting, eye-catching cover. This makes single-focus covers particularly appropriate for magazines heavily dependent on newsstand sales, since large designs coupled with intense colors attract the eye and can be seen from greater distances than smaller, more subdued design patterns.

With this in mind, I sometimes bring along several real magazines to simulate a newsstand display when showing cover comps to a magazine publisher. This has proven helpful in assessing which of the cover designs is likely to be the most effective in the competitive atmosphere of the newsstand.

A *single-topic* magazine, as the name implies, is devoted to the complete exploration of a single area of interest, such as photography, automobiles, electronics, and computers. For many years, the covers commonly used by these magazines consisted of the publication's name in bold type and an array of titles and blurbs to describe the featured articles. In the competitive effort to attract the eye of the consumer, the covers fairly shouted the titles of the articles to be found within the magazine's pages—a technique similar to that used by today's supermarket tabloids.

Faced with increasingly intense newsstand competition over the years, the more progressive magazines recognized the need to project more individuality, eventually leading the publishers to explore the possibilities offered by illustration. As a result, today's single-topic magazine covers feature a broad range of styles that include typographic, photographic, and illustrative techniques.

This cover, one in a series commissioned by Popular Electronics *magazine in their effort to boost newsstand sales, features the lead article, "Home Projection TV," as the subject. The new cover art format made it possible to selectively control design elements to reflect the essential aspects of the lead article. Through generic styling, electronic component parts could be incorporated into the new format without appearing to favor any particular manufacturer. The increased flexibility this provided, which had been impossible under the magazine's previous photography-oriented cover format, led to more imaginative covers and higher sales.*

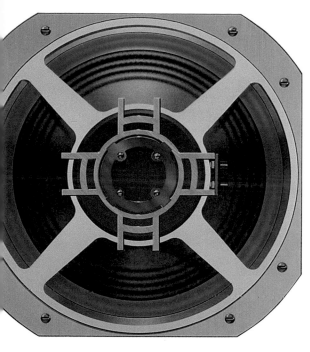

The rear view of a speaker became an effective single-focus cover image for an electronics magazine. By turning the speaker to show its back side, I took advantage of the design symmetry of the driver and frame to create a dynamic, recognizable symbol.

In order for this Popular Electronics cover image to be identifiable as a navigational aid for small boats, I used a ship's wheel and marine flags as graphic devices in the background. Although few readers would realize it, I encoded a "secret message" into the design as well—the flags spell out W-A-R-D, the name of my son.

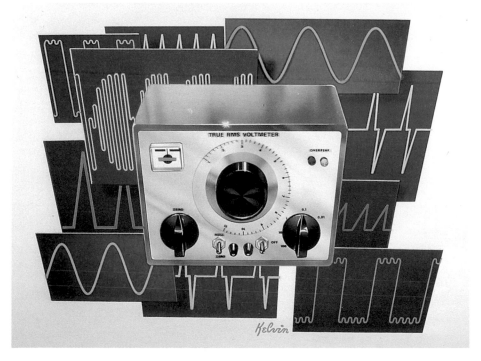

In this example of a generic illustration prepared for a Popular Electronics cover, the wave form in the background and the centrally located voltmeter are immediately recognizable symbols to any electronics buff.

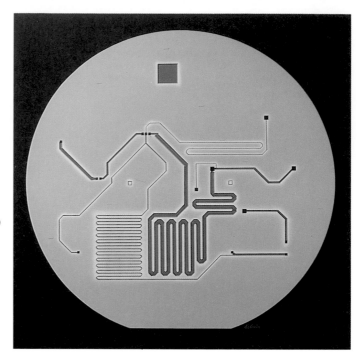

Though the micromechanical device (right) is a real object and squeezed light (below) is a purely theoretical diagram, the geometry and color of both have the contemporary appeal of modern art, creating a pair of eye-catching displays. The squeezed light cover was later chosen for inclusion in a book of great scientific illustrations.

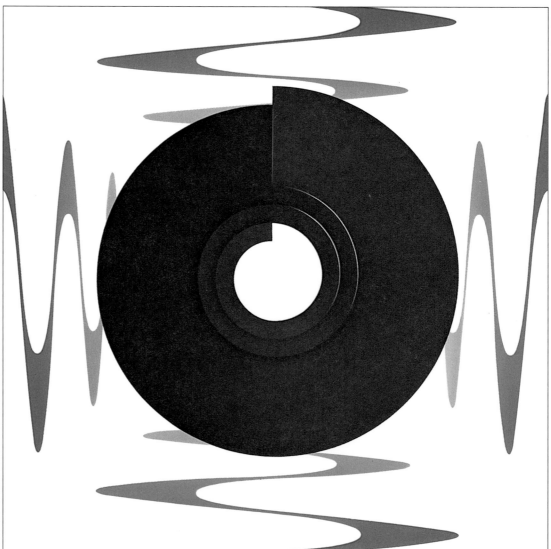

DOUBLE-FOCUS COVERS

Double-focus covers tell two important stories simultaneously, showing the interaction or interrelation of two distinct elements in one illustration. They are well-suited for publications intended for professional audiences, where the material is often complicated, going in many different directions with a great deal of technical detail. Medical and pharmaceutical publications are particularly good candidates for double-focus covers.

Given the complex nature of the material and the subtleties of the story elements' interrelations, a fairly comprehensive understanding of the text is needed before proceeding. When planning a double-focus cover, I first familiarize myself with the reference material and then plan the cover to feature the primary subject of the text, using the secondary information to create an interesting background that will tell its own story. If done well, both elements will be easily understood and be unaffected by any liberties I take regarding the scale of the two components.

Except for the astronaut/moon image on page 181, these illustrations were designed for the covers of *Hospital Practice*, a widely read professional magazine.

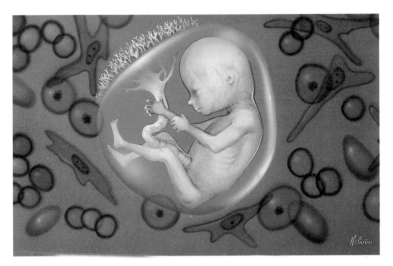

Left: This cover illustration features a portrait of a developing fetus in the foreground against a background of amniotic fluid cells with the potential to affect the fetus's development.

Below: A translucent view of the human nervous system and internal organs is superimposed on a magnified rendering of the neural network.

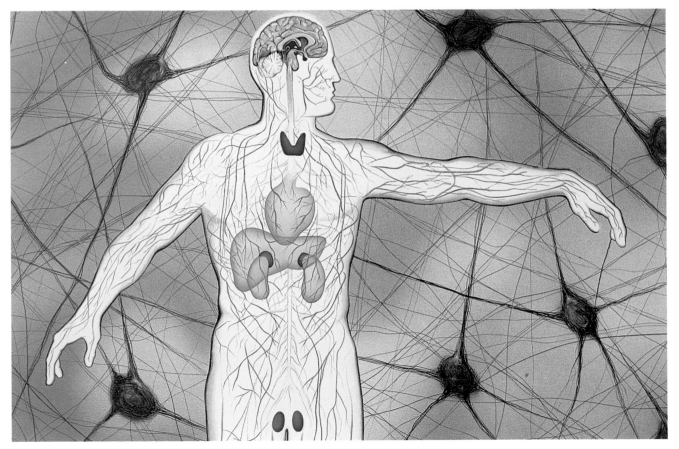

Another double-focus cover image centers on a portrait of a blood-engorged tick carrying the Rocky Mountain spotted fever virus shown against a background of human muscle cells, which are affected by the disease.

Y-shaped antibodies have attached themselves to receptor sites on the surface of a cell, while a single antibody in the foreground shows greater structural detail. Originally designed for Hospital Practice, this painting was also used as a book jacket for a collection of professional articles on immunological diseases.

A somewhat different treatment was given to this Science Digest *cover image commemorating lunar exploration. Designed for newsstand display and a nonprofessional audience, the cover features a partially translucent image of an astronaut floating in front of the lunar surface.*

ADAPTING PATTERNS TO COVER FORMATS

Recomposing the pattern of a simple carbon polymer chain transformed it into a dynamic cover design for *Scientific American* magazine. The polymer chain represents a new plastic product capable of conducting electricity. The basic molecular structure of the chain is rather simple—ionic triplets (represented by the yellow spheres) added to the carbon chain (the black spheres) make up the plastic polymer. A schematic representation of this is shown below.

The assignment presented an interesting design challenge: The format was to be 7 inches (17.8 centimeters) square, and I could tell from the reference material that a simple, single chain would neither fill the space satisfactorily nor do well on the newsstands.

After sketching several routine design variations on a flat plane and discarding all of them, I had an idea for a new approach. Would a bird's-eye view work? Several sketches later, a definite design possibility began to emerge. I constructed a model from plastic rods, photographed it from var-

ious angles, and used the photos as the basis for a perspective drawing. By combining four photographic views of the chain, I reached a very close approximation of my original sketch. Without alteration, the chain had been expanded and given a dramatic dimension.

The final illustration was airbrushed on four-ply Strathmore board, kid finish. To enhance the dramatic effect, the background was painted an ivory black with just a touch of blue. The areas to be occupied by the ionic triplets were masked so the spheres could be rendered on a white surface, which would guarantee the most brilliant color reproduction.

The yellow spheres were rendered starting with a base of lemon yellow, progressing to golden yellow and raw sienna. Note the edges of the yellow spheres in the foreground—by moving the acetate mask slightly after each pass of the airbrush, I achieved a soft, slightly out-of-focus effect. Razor-sharp edges would have created a "pasted-on" look, detracting from the dynamics of the painting.

Schematic representation of the polymer chain's basic structure, as it might be shown in a chemistry book.

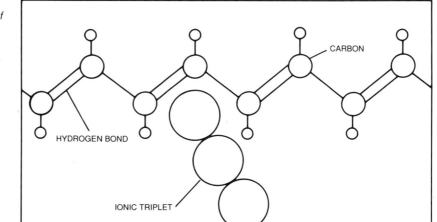

The first three thumbnail sketches show attempts at using just a single polymer chain; in the fourth sketch, four identical chains are combined to create a perspective view of the molecular pattern.

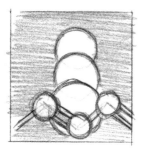
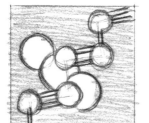
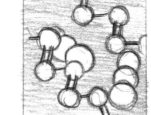

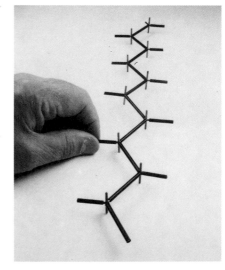

Perspective reference photo of the polymer model. The model was constructed from a molecular model building kit.

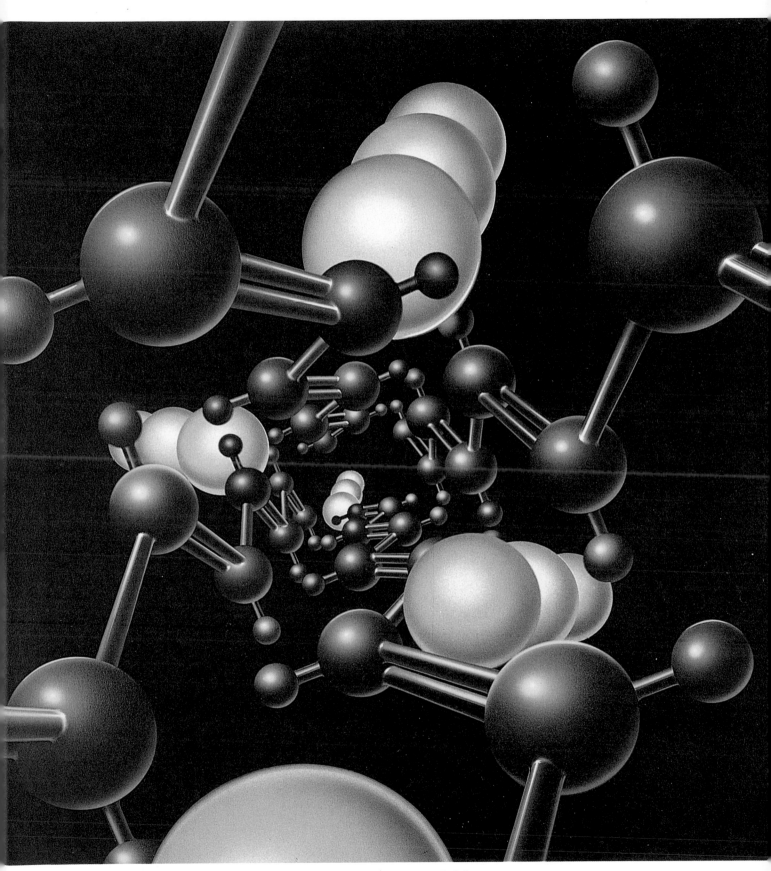

For increased visual impact, the polymer structure was rotated to present an asymmetrical view.

CREATING SOMETHING OUT OF NOTHING

As important as it may be for a scientific illustrator to have outstanding artistic technique, imaginative problem-solving capabilities, diligent research skills, and so on, sometimes the sheer need for resourcefulness outweighs everything else. Some projects, for whatever unusual reasons, simply allow no time for preparation work, offer no choice of reference materials, and basically hang the illustrator out to dry. In these situations, being resourceful is the hallmark of the quick-thinking professional.

In the case of the UFO detector illustration for the cover of *Popular Electronics* magazine, for instance, no information was available. Unusual circumstances required that the cover design be completed prior to the magazine's receipt of the related article. Consequently, nobody had any idea of what this detector actually looked like. In addition, there was a very tight deadline, leaving me no time for preliminary pencil sketches—I was asked to go directly to a final.

In my discussions with the editorial staff, we reasoned that the major components of a UFO detector would have to include a meter to record any UFO signals, tuning knobs, a speaker, and a directional compass. It was left to me to put this all together and design a housing.

This is the kind of assignment to have fun with. My plan for the housing was fairly simple—something between Buck Rodgers and a Geiger counter. To complete the effect, I placed the shadowy form of a UFO, emitting electromagnetic waves, looming menacingly in the background.

The biochips illustration for the cover of *Discover* magazine was another emergency assignment. The feature article originally planned for the cover was being discarded; the substitute article dealt with the possibility of one day developing a computer chip from biological organisms. Computer chips are composed of tens of thousands of electronic switches that can be turned on or off, with information encoded and stored by various combinations of opened and closed circuits. It was proposed that the same "on" and "off" capabilities present in neurons and biological cells could theoretically be used to fabricate a computer chip.

Because of the late substitution, once again there was no time for preliminary sketches. The art director asked me to create a computer chip that looked like it was the product of genetic engineering—not the easiest thing to envision.

I had already produced several hundred drawings over the years relating to bioengineering, so this was a unique opportunity to play scientist. As an icon, I used a generic shape for the base computer chip—a large square with gold connectors on the perimeter and wires leading off from the connectors to other chips and/or computer input devices. The channels and patterns on the surface were designed to be similar but not identical to those on conventional chips. I added a small array of white dots to represent the biological component and included an imposing double helix strand of DNA surrounded by my own concept of the crystalline structure of the biological switches to convey the idea of a biological chip.

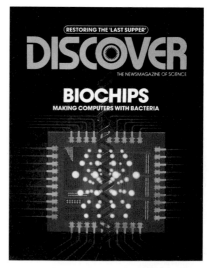

Left: UFO detector of my own design. Strictly speaking, my device has detected just as many verifiable UFOs as any similar "real" device has come up with.

Above and right: The concept of a biological computer chip (background) is conveyed here by a double-helix strand of DNA embedded in a hypothetical crystalline structure (foreground). The viewer's focus is directed through the soft-edged crystalline spheres toward the sharply defined background images.

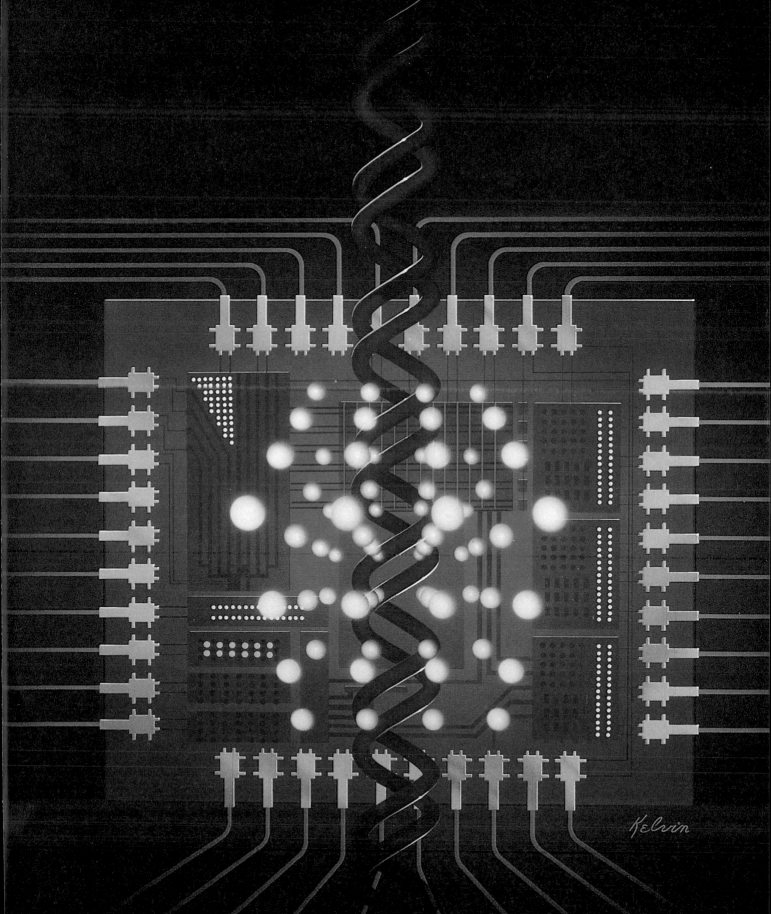

BUSINESS CONSIDERATIONS

People at social gatherings are generally intrigued to learn that I am an artist. Once this discovery is made, the conversation quickly turns to a discussion of art in all its forms. Inevitably, someone assumes that I am involved in the fine arts and eventually asks what my "real" occupation is. When I reply, "I am an artist," the question comes back, "Yes, but what do you do *for a living?*"

Artists engaged in the commercial arts are frequently "finer" artists than those engaged in the fine arts. It is true that commercial artists expect to earn their living from their art while fine artists are willing to work for love, but fine artists price their work by its market value, a luxury that commercial artists cannot afford. Should you decide to go into this business, think of yourself as a businessman/artist, with equal weight to each side of the term. This frame of mind should help you to reap the benefits of sound business practices, as other professional groups do.

CALCULATING OVERHEAD EXPENSES

Determining the cost of your overhead is central to developing a price structure for your work. Heading the list of essential expenses is the cost of acquiring a competent accountant. Take the time to find someone who specializes in small businesses and can guide you in setting up your operation in the most advantageous manner. Be sure to find someone with whom you are comfortable, since this person will be your most important adviser and confidant.

Initially, your accountant will set up your general ledger, help you determine your cost base, and project the tax implications. Self-employed artists pay estimated quarterly taxes, which means an appropriate amount will have to be figured into the cost of every job you accept so that you will be able to meet your tax obligations on time.

Whether you choose to work at home or rent space elsewhere, it costs money to get started. Statistically, there is a high failure rate during the first year for small business operations of all types, usually attributable to undercapitalization. Before leaving a steady job for the challenge of going out on your own, be sure you have sufficient financial resources to help you over the rough spots during the first year of operation. There is a real need for artists interested in developing scientific and technical art businesses—the work is out there, the future looks bright, and the field is not saturated with competitors. But you have to stay in business long enough to establish your niche.

Careful planning will help you to minimize your overhead expenses. The following lists of the most basic office and art supplies can serve as a starting point, and you will surely find a few expenses of your own to add to them. If the lists seem imposing or some of the items seem extraneous, I will only add here that efforts to reduce overhead by using inferior equipment or supplies will eventually prove to be a costly error. There are better areas to economize than here.

ART/STUDIO

Desk
Chair
Desk lamps
Drafting machine
Airbrush
Compressor
Taboret
Art files
Tracing pads
Portfolio
Ellipse guides
Triangles, curves
Enlarging and reducing photocopier
Paper cutter
Fax
Light box
Paints, brushes, palettes
Illustration board
Mechanical pen set
Camera equipment

OFFICE

Stationery
Business cards
Desk
Telephone
Reference books
Calculator
Rolodex
Typewriter
Files and folders
Stamps, pads, etc.

SERVICES

Accountant
Rent
Printing
Federal Express
Health insurance
Electricity
Telephone
Local messenger service
Fire and theft insurance
Travel

HIDDEN EXPENSES

Vacation time
Time lost due to holidays
Time lost due to illness

Start saving those seemingly inconsequential receipts that you usually throw away. Trivial expenses like magazines, garage parking, postage stamps, and assorted nickel-and-dime items are now part of your operating costs—you will be surprised at how they add up. You can use them to decrease your taxes and increase your profit by simply saving the receipts in an envelope for your accountant.

Once your accountant has helped you determine what your weekly operating expenses are likely to be, determine what you must draw in salary in order to maintain your present life-style. Add your weekly operating expenses to your weekly salary, divide by an average work week of 40 hours, and you'll have the hourly rate to charge for your work. Assuming the rate turns out to work well for you and your clients, plan to adjust it annually to keep pace with the rising cost of living.

ESTIMATING A JOB

Learning how to estimate a job is a skill that comes with experience. If you overestimate the cost to the client, you will not get the assignment; if you underestimate the time it will take you to do the work, you cheat yourself.

There are artists who price strictly by size and art style—a full-page, full-color painting costs a certain specified amount, a half-page line drawing is a different specified amount, and so on. This never has made much sense to me, since two illustrations of the same size and style are not necessarily comparable in other ways—one might take only a few hours while the other might require weeks of carefully researched, detailed work.

Most estimates are given in two stages. The first stage occurs when the client first contacts you about an assignment. The initial telephone conversation, which generally includes basic information about the project and the kind of budget and deadlines the client is working with, should at least tell you whether you can afford to be interested in the job. Sometimes the client can arrange to send you some in-house preliminary sketches to estimate from, which makes your job a little easier; otherwise, you can set up an assignment conference (see Chapter 5). By the end of the conference, you should have an idea of how much research you will have to do on your own, whether you will have to build a model or take on-site photographs, and if there will be technical people to provide you with the resource material you might need.

With this information factored in, you can estimate approximately how many hours the job should take. This takes practice, but you will learn from your mistakes. Your hourly rate multiplied by the number of hours needed for the job is the fee you should charge.

If the client agrees to the fee, there is no problem; if the client gasps and goes pale, ask what he or she feels the job should cost. To close the gap between your rate and the client's budget, be prepared to offer alternate ways of approaching the assignment. For example, perhaps the illustration can be shown in line instead of in tone, or from a different view. There are dozens of ways to design art in order to bring it into a budget, so be prepared to think creatively.

When faced with these situations, I explain to the client how I can simplify the work yet still produce an attractive, well-designed piece of art that will meet all the requirements even though it contains fewer details. A quick, rough sketch of what I have in mind usually helps. For example, if a bolt really has 60 threads but I only put 30 in the drawing, does it make a difference? To the purist it might, but to most people it will still be a bolt.

If I succeed in convincing the client that modifying the art will not cause the quality to suffer, we can usually come to an agreement; if the client insists on the original concept at a lower price than I can afford, I decline the job as graciously as possible.

Sometimes, however, as your fledgling business begins to take hold, there will be instances when you will accept an assignment for less money than the job actually calls for. The circumstances might include the prospect of developing an account that could generate future assignments, the possibility of making good career contacts, and so on. In these situations, the benefits outweigh the reduced revenue, and it makes sense to take the job at a lower rate.

If such a job arises, make a point of listening and responding to the client's needs, be astute in asking the right questions, and be sure to deliver a quality job on time. If you work with a sensitive approach, you should have no problem getting your correct fee on the next assignment.

In a case where you believe all aspects of the job were handled well but the client still finds your next estimate too high, be prepared to say, "Thanks, but no thanks." Of course, if this scenario repeats itself too often, you probably should review your overhead and salary needs—you may really be asking for too much money. Through trial and error and regular evaluation of your business procedures, you will ultimately arrive at a rate that you and your clients can live with. In short, the market will bring you into proper alignment.

This brings me to one of the more difficult questions you may have to face during that important first year: Should you routinely accept work that brings in enough money to cover your office expenses but is not enough to cover your salary? My answer would have to be no. Do not make the mistake of thinking that any work that pays your bills is better than no work at all—this is no better than working for free. As difficult as it might be to accept, you probably would be better off throwing in the towel and finding a job with regular hours and benefits.

PURCHASE ORDERS AND CONTRACTS

Request a purchase order or contract when accepting any assignment. It should include a complete description of the job, delivery dates for pencil sketches and final art, the total purchase price, and the date when you will be paid.

Be sure the agreement states whether *all* rights to the art are being purchased or only *first-reproduction* rights. When you agree to sell all rights, the buyer owns the original art and may use or reproduce the art as often or in any manner he or she chooses, and may alter or crop the art for such uses; when only first-reproduction rights are sold, the buyer may use the art only once and the ownership of the original art remains with the artist who may sell secondary publishing rights to other clients. With secondary rights the art may not be traced, copied or altered in any manner without express permission from the artist.

The type of rights to sell a client has to be decided on a case-by-case basis. However, the price of the art should reflect the terms of the agreement—obviously, if you agree to sell all rights, you should charge more.

SOLICITING WORK

Getting started as a scientific illustrator, like getting started in anything else, calls for legwork and persistence. Try calling companies specializing in the servicing or manufacturing of technical or scientific products and find out who does their brochures, annual reports, and high-tech advertising. Talk to college science department chairs, who can direct you to the publishing companies from which they buy their books. Check out the science magazines at your corner newsstand and contact their art directors. And while you're there, don't overlook the local paper and its want ad listings. A more expensive but noteworthy approach is to solicit through one of the popular artist and illustrator directories, which will sell you quarter-, half-, or full-page display space. Most art directors keep desk copies of these directories and use them to locate artists for special projects.

Make appointments to show your portfolio and compile a mailing list. One or two responses more than pays for a mailing. Utilize reference books, like *Literary Marketplace* (Bowker Publishers), to target your market. Even local phone directories offer good sources of publisher and advertising agencies with the potential of becoming clients. Picking up the phone and learning to get through to the art director is a valuable experience in itself, and when you catch an art director with a few minutes to spare, you can sometimes get worthwhile advice, new leads, and, if you're lucky, an appointment to show your portfolio.

Explore the possibility of acquiring work in areas often overlooked because they are not considered as glamorous as magazine and book work. Charts and graphs, annual reports, and technical catalogs can become lucrative sources of bread-and-butter work that can keep you going until you stabilize your income.

PREPARING AND PRESENTING YOUR PORTFOLIO

Your portfolio reveals everything there is to know about you as an artist and as a professional. To the prospective employer or client, it serves as a clear indicator of where you have been, where you are going, and what you are worth. It is the ultimate barometer of your career as an illustrator.

Choosing what to include in a portfolio is always a difficult task, particularly for the recent art school graduate who has not yet had the opportunity to build a body of professional-level work. If you are in this situation, most of your work will have received two types of treatment: advice and criticism from teachers and fellow students, and love and adoration from family and friends. What it will not have been exposed to, however, are the demands and expectations of the marketplace.

With this in mind, I suggest a rather atypical approach: *start fresh.* After all, student work is student work; you can't expect it to bring you a career. In preparation for becoming a scientific illustrator, burn, hide, or store everything you did as a student and devote a few hours per day to new portfolio pieces.

Keep the subject matter simple—it is better to do a simple job very well than a complicated project poorly. Target your first portfolio toward a realistic market. Concentrate on showcasing strong rendering skills. Charts and graphs and carefully rendered line drawings will go a long way toward securing a staff position in a technical art studio or acquiring free-lance assignments. With time and experience, your portfolio will reflect more challenging work and begin to attract the more creative assignments.

It is generally assumed (fairly or not) that artists engaged in scientific illustration maintain their portfolios as models of logic, neatness, detail, and accuracy. To this end, and for the most effective presentation, I suggest using a portfolio with acetate pages. The size of the portfolio (or *book,* as it is commonly called) depends on the size of the artwork you plan to display; in general, choose one as lightweight and small as your needs permit. The first page should be reserved for a brief résumé summarizing your formal training, work experience, and goals.

Show a well-rounded background by including sample illustrations for the various scientific and technical areas in which you are prepared to accept assignments. Each page should display an attractive grouping of work specifically relating to a given subject, such as astronomy, biology, geology, and so on. Use both sides of each page, but resist the temptation to include too many pieces—a selection of four or five well-done renderings on any one topic is quite enough.

A professional portfolio should always reflect current work. A favorite piece from several years ago, despite its virtues, begins looking outdated if it remains in your portfolio for too long. Moreover, it may contain information that has long since become inaccurate or obsolete, or it simply may not represent your current skills, your growth, and the maturation of artistic style reflected in your more recent work.

When you are asked to leave your portfolio for review, remember that a well-assembled book can speak for itself. If you are present when your portfolio is being reviewed, be prepared to answer technical questions about the art. This is an opportunity to show your knowledge of the subject, but don't overdo it—be brief and to the point. If you have included less than perfect samples of your work for any reason—perhaps you have a bit of trouble with biological subjects but opted to include a substandard biological piece rather than none at all—do not volunteer an explanation for having done so. If it requires that kind of discussion, it does not belong there to start with.

An art director should be able to review your work easily and quickly. The last-minute inclusion of illustrations or matted pieces distracts from the clean, organized appearance your portfolio should project. If you have slides or samples of published work that you can leave, that is always a good move. Prepare these samples in advance in a separate folder and include a copy of your résumé.

In discussing a career in illustrating for science, I have attempted to go beyond the mechanics and techniques involved, and to express the sense of satisfaction that goes with the territory. These rewards come from much more than just being a successful illustrator—they come from being able to contribute to the new ideas that will take us into the next century and, when luck and circumstance permit, from the opportunity to stand for just a moment with the men and women on the frontiers of science.

ILLUSTRATION CREDITS

The author and publisher gratefully acknowledge the following sources, which have given their permission for images to appear in the book:

INDEX